RUN, SWIM, THROW, CHEAT

Chris Cooper is a distinguished biochemist with over 20 years research and teaching experience. He is Professor and Director of the Centre for Sports and Exercise Science at the University of Essex, where his research interests include developing artificial blood to replace red cell transfusions. He is a regular broadcaster on issues relating to sport and science.

RUN, SWIM, THROW, CHEAT

The science behind drugs in sport

Chris Cooper

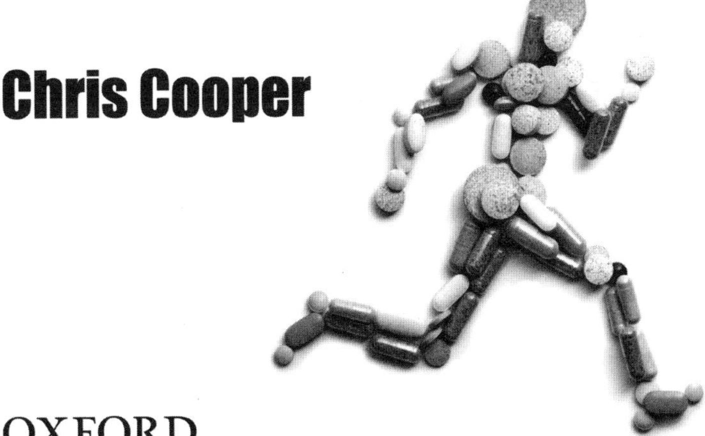

OXFORD
UNIVERSITY PRESS

OXFORD
UNIVERSITY PRESS

Great Clarendon Street, Oxford, OX2 6DP,
United Kingdom

Oxford University Press is a department of the University of Oxford.
It furthers the University's objective of excellence in research, scholarship,
and education by publishing worldwide. Oxford is a registered trade mark of
Oxford University Press in the UK and in certain other countries

First Edition published in 2012
First published in paperback 2013
Impression: 1

Published in the United States of America by Oxford University Press
198 Madison Avenue, New York, NY 10016, United States of America

British Library Cataloguing in Publication Data
Data available

ISBN 978–0–19–958146–7 (hbk.)
ISBN 978–0–19–967878–5 (pbk.)

Printed in Great Britain on
acid-free paper by
Clays Ltd, St Ives plc

CONTENTS

ILLUSTRATIONS

ACKNOWLEDGEMENTS

The idea for this book arose from discussions with students and staff at the Centre for Sports and Exercise Science at the University of Essex. I particularly thank Jerry Shearman, Martin Sellens, and Mike Wilson who encouraged me to think of applying my biochemical ideas to sports science. I am also grateful for the support of the many scientists—at Essex and elsewhere—who suggested helpful improvements to the content of this book. These include Ralph Beneke, Yagesh Bhambani, Stephen Harridge, Gareth Jones, Hugh Montgomery, Ceri Nicholas, Peter Rasmussen, Kevin Tipton, Niels Vollaard, Anna Wittekind and John Wyatt.

This has been my first book targeted at a general readership. I am indebted to Jules Pretty for supporting me in this endeavour and contributing valuable advice both on the scientific content and, especially, on the art of popular science writing. My wife, Helen Cooper, provided the insights—as well as the aggressive proofreading—of an intelligent non-scientist.

I especially wish to acknowledge the UK's Engineering and Physical Sciences Research Council (EPSRC) for providing the support of their innovative Senior Media Fellowship programme; this paid for relief from my formal academic duties, providing the free time to dedicate to writing. At Oxford University Press, Emma Marchant and Erica Martin assisted with technical details of the book whilst Latha Menon provided valuable advice on content.

Finally writing a book is a time consuming and emotional commitment. I could not have succeeded without the support of my family especially my mother Maria, my wife Helen and my two children Lauren and Alex.

PREFACE TO THE PAPERBACK EDITION: 2012 UPDATE

For all British sports fans, 2012 will be seen as a once in a lifetime experience. There was the first UK Tour de France winner, the first male Grand Slam tennis champion since 1936 and a British golfer headed the money list for both the US and European tours. Above all there was the London 2012 Olympics and Paralympics, complete with a skydiving Queen and record medal hauls for the home nation.

As is usual, the anti-doping preparations for the Olympic games were hyped as the best ever and, as usual, very few positive tests were reported at the games. I had the dubious privilege of being in the Olympic stadium the only night a medal was won by someone who was subsequently banned for doping. Nadzeya Ostapchuk, the Belarusian shot putter, won gold, beating the reigning champion Valerie Adams, much to the chagrin of the New Zealand journalists sitting around me. Ostapchuk's positive test for the anabolic steroid metenolone resulted in her being stripped of her medal by the International Olympic Committee (IOC).

Some might say that the paucity of positive tests meant that only the "dopey" dopers were caught, most athletes taking banned substances having stopped prior to the games to avoid detection. But a closer look reveals a more complex picture. Over 100 athletes were banned from even going to London, having failed targeted testing in the period prior to the games. Increasing use of intelligence as to who might be doping has become a feature of anti-doping strategies, as is the re-testing of old samples using more modern analytical techniques. Immediately prior to the London 2012 Olympic games, three runners were banned from competing due to new tests on urine samples taken a year earlier at the Daewoo 2011 World Athletics championships.

Blood and urine samples given at the Olympics can be stored for eight years after the games end. In addition, at the London games new techniques were pioneered to measure signals from as many substances as possible, rather than just the 200 or so for which a validated doping test is in place. Should suspicions arise about an undetectable drug, this database can be screened for the appropriate signal and any abnormal samples tested again. The eight-year rule is no idle threat. Before the start of the London 2012 games, there was a rush to re-test all the Athens 2004 Olympics samples using the latest, more sophisticated methods. This resulted in four athletes losing their medals. One was Yuriy Bilonog who "won" the men's shot put. He therefore joined the original winner of the women's event, Irina Korzhanenko, in being stripped by the IOC of the gold medal due to a positive test for an anabolic steroid (though neither athlete subsequently admitted doping).

As the only event held in the original ancient Olympia stadium, the shot put competition was to have been a highlight of the Athens games. We will never know whether the original Greek Olympians would have been perturbed that both gold medalists were found to have ingested banned substances. For, although the ancient Greeks did indeed punish cheating by disqualification and heavy fines (the latter being used to erect statues of Zeus), as far as we know they did not consider any of the nutritional aids or potions taken by the athletes to be cheating (see Chapter 1).

All these Olympic moments paled into insignificance compared to what hit the world of professional cycling in 2012. In October, the United States Anti-Doping Agency (USADA) published over 1,000 pages of evidence that, in their words, showed that "the achievements of the USPS/Discovery Channel Pro Cycling Team, including those of Lance Armstrong as its leader, were accomplished through a massive team doping scheme, more extensive than any previously revealed in professional sports history" [1]. In an interview with Oprah Winfrey on January 18, 2013, Lance Armstrong finally admitted doping during the period of his Tour de France wins (1999–2005).

This book deliberately focuses on the science, not the individual. There is no attempt to expose, directly or by implication, any athlete who has not been convicted of a doping offence. Therefore in the current version there is only passing reference to Lance Armstrong. Yet all of the methods he was convicted of using are well documented in its various chapters. One thing I do state is that he was able to use oxygen efficiently without producing excess lactic acid. We know that much of this was due to methods such as EPO and blood doping that enhance oxygen delivery to the muscle (outlined in Chapter 4 of the book). What the USADA report also reveals is the combination of high-tech and low-tech methods Armstrong and his team used to evade detection (see Chapter 10). The high-tech method involved doping with EPO and steroids at the optimum moment, and via the best delivery method, to narrow the window of time when a sample could test positive. The low-tech method involved not answering the door, or hiding if a doping control officer arrived within that detection window.

We will never know exactly how good Armstrong was. It is clear that he was not the only cyclist doping in the period of his Tour triumphs. Three possibilities present themselves. He was naturally a better cyclist and his doping merely levelled the playing field, at least amongst those who chose to dope; his body's physiology was more suited to being enhanced by doping; or he had a more effective doping regime.

The USADA case against Armstrong was mostly based on evidence given by his teammates, who claim to have seen him doping. However, there was some scientific supportive evidence, though not of the type that could result in a conviction in its own right. The first example came from the 1999 Tour de France. The urine samples collected in this Tour were later reanalysed to develop a new modified test for EPO doping. The results were meant to be anonymous, but a journalist found a way to break this code. Armstrong's samples accounted for 46% of all the EPO positive samples found, suggesting he might have been more aggressively doping at this time than his rivals. The second piece of evidence came from blood samples taken from Armstrong during his comeback tours in 2009 and 2010. In many of Armstrong's blood samples there was a dramatic reduction in the number of young red blood cells. These young cells, called reticulocytes, are formed when you are in the process of making new mature red cells. If you transfuse blood to supramaximal levels, your body sends a signal to stop making new red cells. Therefore there is a large decrease in the number of reticulocytes. Armstrong showed exactly this effect on a number of occasions. This is the relevant quote from the report "Prof. Gore concluded that the approximate likelihood of Armstrong's seven suppressed reticulocyte values during the 2009 and 2010 Tours de France occurring naturally was less than one in a million".

The political repercussions of the Armstrong revelations for the sport of professional cycling will take some time to resolve. Yet we should not lose sight of the fact that most of the events described occurred over seven years ago. Many observers would support the view that there is less doping in cycling at present, certainly of the orchestrated type facilitated by the teams in the 1990s and 2000s. Cycling perforce has instigated new anti-doping regimes that are much more stringent than in other sports. One is the biological passport, where a number of the athlete's blood parameters are monitored over a period of time to determine whether the red blood cell levels have been manipulated artificially. An abnormal spike in the readings can trigger a penalty, even in the absence of a specific measurement of a banned substance. This tool has since been taken up by track and field athletics; six runners were sanctioned immediately prior to the London Olympics. Armstrong himself said, in an interview with Oprah Winfrey, that the passport was the reason it would be harder for him to repeat his doping today. As I state in the book (Chapter 10) there are differing views about how easy it is to "fool" the passport. But there are no doubts that it is here to stay, at least for sports that can afford the extra expense to implement it.

The year 2012 provided another first in doping with veteran triathlete Kevin Moats testing positive for testosterone whilst competing in the 55–59 year age category. Despite providing evidence that his doctor had prescribed him testosterone replacement therapy, he was banned for one year by the World Triathlon Corporation. Male hormones naturally decrease with age. With testosterone supplementation increasingly being offered as a medical treatment for older men, this area is likely to be of growing concern to athletes and sporting bodies.

What does the future hold for new drugs? The sprinter's dream (see Chapter 6) of being able to increase muscle mass and power without the sexual side effects of anabolic steroids, took a step closer as a Selective Androgen Receptor Modulator (SARM) posted promising clinical trial results, improving muscle mass in elderly patients and post menopausal women without apparent side effects [2]. For athletes in endurance sports, a hypoxia-inducible factor activator stabilizer (see Chapter 4) was shown to increase the red blood cell content of kidney dialysis patients [3], providing a potentially undetectable alternative to EPO for unscrupulous athletes. There is no doubt that every year will see new drugs coming to the market with potential sports doping benefits. And every year the anti doping agencies will struggle to develop a suitable test.

Experiencing a home Olympics did drive home to me the ability of sport to lift the spirits of a cynical nation in the midst of a seemingly endless economic depression. Even four months after the closing ceremony, 80% of British people surveyed thought the games were worth the £9 billion price tag. Upholding this "spirit of sport" is central to the World Anti-Doping Agencies policies. It underpins the policy of banning recreational drugs even if they are not performance enhancing. If you read Chapter 9 you will see that I was rather dismissive of the idealism associated with such a vague concept as the spirit of sport. So I was somewhat surprised that in a straw poll the vast majority of my current students agreed with the view that elite athletes should be role models for society and that recreational drugs should remain on the banned list. However, having felt the Olympic movement first hand, I do now have more sympathy for the transformational ability of sport and the resultant responsibilities imposed on its stars.

Following the book's publication, I have had the privilege of many discussions with members of the public, representatives of sporting bodies and elite athletes. This brought into sharp focus for me the effect of doping on the spectacle of sport and on the lives of those athletes who don't dope. Like match fixing, doping is hidden cheating. This is one reason the penalties are so severe. Many elite athletes are accused of cheating with no evidence other than that they can run fast, swim quick or throw long. Indeed, at the Olympics, more outrage was vented by some media commentators on people who performed beyond expectations, and were assumed to be cheating, than on those few who were actually caught doping. The Lance Armstrong case has only added to this problem, for he performed amazing deeds, yet never failed a drugs test. The shadow of doping taints clean athletes almost as much as the guilty. As the book shows, you cannot wave a magic wand to solve this problem, but knowledge of the science can surely assist in formulating practical solutions.

If you want more detail on these recent stories or wish to keep up with the constantly developing science behind the latest news about doping in sport, the book's blog can be found at www.runswimthrowcheat.com

Prologue: A tale of two races

The 1988 Men's 100 m Olympic Final has been called the most corrupt race ever.[1] It took everyone's breath away. One man ran a tenth of a second faster than any person had ever done. The world record was smashed. The world celebrated, for a day. And then the new myth fell apart. Ben Johnson, hero of all Canada, was a cheat. Though some of the offences were minor compared to his, subsequent events implicated six of the eight sprinters in that race in activities that—at one time or another—would have them banned from winning Olympic gold medals.

I remember the race vividly. I was researching my PhD in Canada at the time, but was on holiday back in the UK. Two Canadians and a Briton were lined up in the final. I stayed up until four o'clock in the morning. Although it could be argued that I had dual loyalty between the Commonwealth allies, my main desire was in rooting for the underdog. Could someone beat the seemingly all-powerful US team—in this case exemplified by the not-quite-so-humble current Olympic champion and former world record holder Carl Lewis? Ben Johnson the current world record holder and

Canadian hero seemed a likely prospect, but he had barely survived the semi-final. And yet, one man won convincingly (see Figure 1).

Johnson was first in an astonishing time of 9.79 s—more than 0.1 s faster than any other man had ever run and a time set whilst celebrating victory before reaching the line. But what happened two days later shocked the sporting world, and especially Canada. Johnson tested positive for the banned anabolic steroid stanozolol, the gold medal was removed, and his career collapsed. He tested positive for steroids a second time during an abortive comeback in 1993 and was banned for life by the International Association of Athletics Federations (IAAF).

Second place went to Carl Lewis (USA) who was awarded the gold medal when Ben Johnson was disqualified. However, at the 1988 US Olympic trials Lewis tested positive for the banned stimulants pseudoephedrine, ephedrine and phenylpropanolamine. At the time the US Olympic Committee (USOC) policy was not to ban athletes for stimulant use unless it was demonstrated that there was a deliberate attempt to affect performance. Inadvertent use of herbal supplements or over-the-counter cold remedies would not result in a ban (a decision upheld by an IAAF investigation in 2003). The same defence has not worked in other sports illustrating the anomalies of anti-doping systems. For example the 16-year old Romanian gymnast, Andreea Raducan, was stripped of her gold medal at the 2000 Sydney Olympics when her coach gave her a pseudoephedrine-containing Nurofen pill as a cold therapy.

In third place came Linford Christie (UK), who in the next Olympics (Barcelona, 1992) was to become—and still is—the oldest man ever to win the Olympic 100 m title. But after the Seoul final, Christie tested positive for the same stimulant for which Lewis had tested positive, namely pseudoephedrine. By a vote of eleven to ten the International Olympic Committee's disciplinary committee accepted his response that he had taken it inadvertently in ginseng tea. However, Christie was caught when he came out of semi-retirement to race in Dortmund, Germany at the age of 39. He tested positive for the steroid nandrolone and the IAAF banned him for two years. Christie has always maintained his innocence of this charge.

Figure 1 The 1988 Olympic Men's 100 m race in Seoul.

Subsequent events showed that six out of the eight athletes used banned substances in their careers.

Calvin Smith (USA) was fourth on the day and clean then and since. Should he have been given the gold medal? He certainly thinks so. 'I should have been the gold medallist,... If all the positives (tests) are true, I should have won.'[2]

Fifth place went to Dennis Mitchell (USA). Ten years later Mitchell was banned for two years by the IAAF for an excess of testosterone; he also admitted to being injected with human growth hormone.[3] It is not known whether he was on performance-enhancing drugs at the time of the Seoul race.

Sixth place went to Robson De Silva (Brazil), a clean athlete who had the consolation of a bronze medal in the 200 m at the same Olympics. Yet in gold and silver medal positions in that race were Joe DeLoach and Carl Lewis, both beneficiaries of the US Olympic Committee's policy on disregarding positive tests triggered by inadvertent stimulant use during the Olympic trials. If no justifications were accepted for stimulant use De Silva could have had gold and silver at these Olympics, but came away with only bronze.

Desai Williams (Canada) ran in seventh. He admitted taking steroids from September 1987 to January 1988.[4] Although Williams always denied subsequent use[5] the Dubin inquiry concluded that it was 'satisfied that Mr. Williams was using anabolic steroids not only in the fall of 1987 but also during the spring and summer of 1988 prior to the Seoul Olympics'.[4] Williams was banned for life from receiving Canadian sports funding although this ban was lifted in 2010 due to his subsequent positive behaviour.[5]

Eighth place went to Ray Stewart (Jamaica). There is no evidence that he took drugs himself in his running career, but in 2010 he received a lifetime ban from the US Anti-Doping Agency (USADA) for providing a range of performance enhancing drugs to the athletes he coached.

The aftermath of Seoul

Johnson's ban for taking the steroid stanozolol led directly to the Dubin 'Inquiry Into the Use of Drugs and Banned Practices Intended to Increase Athletic Performance'.[4] This move by the Canadian government was the first real attempt by Western sport to publicise openly what many knew was happening in private. A host of Canadian athletes and coaches revealed the dirty tricks of their trade. It is illuminating to listen to these.[6] At the inquiry it was noted that the standard procedure after an athlete is caught doping was to 'deny, deny, deny'. For ninety-one days in a courtroom in Toronto this 'standard procedure' was ignored as testimony from forty-eight steroid users—in sports as wide-ranging as track-and field, weightlifting, bobsleigh, wrestling and football—filled 14,617 pages of transcript. This volume of data was later matched when the Stasi files illustrating the extent of the East German doping programme in the 1970s and 1980s were revealed, following the fall of the Berlin Wall. Over 2,000 athletes, including many children, were revealed to have been systematically doped. So did these events, outlining the extent of doping in the West and the East have an effect on sporting practice? Absolutely not if you listen to the athletes themselves.

It is not surprising that athletes sometimes hint that other competitors have an unfair advantage. But it is rare to name specific rivals. Whilst sprinters and anabolic steroids featured in the Dubin inquiry and the Stasi files, the early years of the twenty-first century saw several prominent UK athletes voicing their concerns about the use of a drug made infamous by the Tour de France cycling race—the blood booster EPO that helps athletes use oxygen more efficiently in long distance events.

The extent of the criticism of fellow athletes was surprising. In the 2001 World Athletic Championships, Paula Radcliffe, Hayley Tullett and Kathy Butler unfurled a banner that read 'EPO cheats out' as a rival athlete Olga Yegorova ran in her heats. Then Britain's Kelly Holmes accused Slovenia's Jolanda Ceplak after she lost to her in the women's 800 m race at the 2002 European Championships.[7] When asked about Ceplak's win Holmes said, 'take your own guess [how she did it]. I know I do it fairly . . .'. Ceplak's reply to Holmes was right out of the athlete's 'standard procedure': 'I think she [Kelly Holmes] needs to make me a big apology. She said she did it fairly, but I did too.'

So did the weight of the legal inquiries, captured Stasi files and the voluble protests of the athletes themselves have a long term effect? Let's fast forward to 2005 to find a race that is less famous than Ben Johnson's sprint but is every bit as indicative of the powerful hold that drugs have in sport.

The most corrupt race ever?

In 2004 the UK's Observer newspaper claimed that Ben Johnson's 1988 victory was the most corrupt race ever. This title was short lived. In 2005 the World Championship women's 1500 m race was run in Helsinki. Here all of the first five athletes to cross the finishing line were found guilty by the IAAF of doping offences in their subsequent careers (see Figure 2).

Figure 2 The 2005 World Championship 1500 m race in Helsinki.
Subsequent events showed that the first five athletes to cross the line were found guilty of doping offences in their careers. Although never admitting to their guilt the IAAF bans were never overturned.

First place went to Russia's Tatyana Tomashova. Curiously, her fall from grace came because she was too good at being tested for drugs. Normally out-of-competition testers have to hunt for an athlete once they have given them their statutory one-hour notice. But they always found Tomashova ready and waiting exactly when requested. The anti-doping officials suspected she was being tipped off and, though this was never proven, they were concerned enough to check previous samples. Forensic tests were used to measure the small amount of DNA in the urine samples; these showed conclusively that the samples given in a 2007 out-of-competition test in Russia did not match those given at the World Championships in Japan the same year. The DNA was from different people. She was banned for two years in 2008.

Second place would have gone to Yuliya Chizhenko-Fomenko (Russia), but she was disqualified for obstructing another athlete. She was also banned in 2008 for the same urine tampering offence as Tomashova.

Third place (later upgraded to the silver medal) was awarded to the Russian Olga Yegorova. She too provided false urine in 2007 and was subsequently banned for two years. However, her history gives us a hint as to what may have been concealed in the missing urine samples. In 2001 she had tested positive for EPO by a laboratory in France, the result that inspired the protests from Paula Radcliffe and her teammates. However, at the time the rule was that a direct urine measurement for EPO had to be backed up by a blood test. This confirmatory test was never performed so no further action was taken against Yegerova at that time.

Fourth place went to Bouchra Ghezielle of France, who was awarded the bronze medal after Chizhenko-Fomenko's disqualification. Ghezielle was later found guilty of doping with EPO in 2008 and given a four-year ban.

Fifth went to another Russian, Yelena Soboleva who was another found guilty of urine tampering in 2007.

Sixth place went to the first 'clean' athlete in the race: Jamal Maryam Yusuf of Bahrain. Yusuf was the unfortunate athlete who was physically obstructed by Chizhenko-Fomenko during the race. Justice on both accounts (doping and obstructing) was eventually done when Yusuf won the gold medal at the next World Championships in Osaka in 2007, beating the Russian Soboleva into second place.

Finally what of Jolanda Ceplak, the athlete accused by Kelly Holmes in 2001? In 2004 Holmes had her revenge on the track, winning the gold medal at the Olympic Games and beating Ceplak into third place. Prior to that Holmes and Ceplak appeared to have made up. Ceplak acted as a pacemaker for Holmes's attempt to run a new UK indoor 800 m record in 2003 and later helping by lecturing to students at 'On Camp with Kelly', Holmes's initiative to identify talented young runners.[8] But then in 2007 Ceplak failed a test for EPO; she was consequently banned from competition for two years by the IAAF. Had she been cheating all the time, or only turned to drugs when her career started to fall apart? Or, as she has always protested her innocence, was something wrong with the test?

The future of drugs?

There is a famous survey by Dr Robert Goldman that claims that over fifty per cent of athletes would take a drug that guaranteed them unlimited, undetectable sporting victories for five years, even if it was followed by instant death.[9] This prologue illustrates the ubiquity of drug use in sport, the extremes some people will go to succeed and the difficulty the authorities have in eradicating the problem. Even a relatively small US company—the Bay Area Laboratory Co-Operative (BALCO)—was able to market drugs to over twenty top athletes in a wide range of different sports, including Olympic sprint champions and record-breaking baseball home run hitters. The fall out from this scandal in 2008, and the resulting US grand Jury investigations, have had at least as much impact in sport as the East German doping programme or the Ben Johnson affair.

It will become clear in this book that we have only scratched the surface of possible artificial enhancements. The science, especially of genetics, has expanded quicker than the 'drug cheats' have been able to keep up. The number of performance enhancing drugs that have ever been used to date is a tiny fraction of the total possible compounds that we can conceive of producing. We know that there is a lot we don't know.

We know so little because we spend so little. In nearly all cases the limiting factor in our knowledge is lack of money. Despite what many may think, in global terms sport is poorly funded. The market for medical and recreational drugs dwarfs the market for those used for performance enhancement. Only where the requirements to compete at sport overlap with a medical need are dopers able to move quickly, piggybacking on new pharmaceutical discoveries to enhance performance.

There are clear parallels between drugs in sport and drugs in society. Both are here to stay. We may not like them; we might rage against them; some politicians even think it worth fighting a war against them. But it is a war that cannot be won. This does not mean there should be no efforts to limit drug use—giving up could be worse. But no one should think there will ever be a time when humans can be completely prevented from using chemistry to enhance their sports performance.

Whether active scientists, athletes, or armchair fans we should empower ourselves to enter the debate about drugs in sport informed by the latest scientific findings, rather than by the words of politicians or disgraced coaches. In this world of sporting drugs and cheats there is much talk and policy that is arbitrary and unscientific; it need not be so.

1 Introduction

'If anyone competes as an athlete, he does not receive the victor's crown unless he competes according to the rules'
(The Second Letter of St Paul to Timothy 2:5)

What's in a name?

Run, Swim, Throw, Cheat is a scientific exploration of the role of drugs in sport. Why do athletes take them and risk their health? Do they work? And there are long-term implications for our views of sport in society. For what will a chemically-enhanced future sportsperson look like? Some of the answers to these questions lie in the social and political spheres. But even then they are influenced by the science. And the science is moving forward rapidly.

What do we think of drugs and sport? This is not a question with a simple answer across all historical time and cultural groups. Many people's views are coloured by a rather hopeful view of sport as an honest competition that must be played on a level playing field. Drug-takers must be cheats; rules are being broken. I sympathise with these views. Although perhaps I am seduced by my nationality. As Roland Renson says[1] 'Modern sport originated in Great Britain as a cultural product of modernity, emphasizing equality and competition. Fair play was the moral creed of this new sporting ethos, created by nineteenth century upper and upper-middle class Englishmen'.

Sport has not completely abandoned this idealistic view. In the language of sporting chemicals there is a careful division in nomenclature between the accepted 'ergogenic aids' and the unacceptable performance-enhancing 'drugs'. Yet the distinction between these two definitions is complex and variable. The authorities have been known on more than one occasion, notably with regards to stimulants, to transfer molecules back and forth between categories. One person's ergogenic aid is another's drug; things are not as black and white as a simple reading of Renson might suggest.

In our present historical times, drugs in sport have existed in the context of both national and individual competition. East Germany proclaimed the superiority of its political system by drug-induced gold medals; in the West capitalism led to more diverse, but not necessarily less effective, systems of delivering similar outcomes. For every communist East European Petra Schneider racing over ten seconds clear of British heroine Sharon Davies in Olympic 400 m swimming, there was a capitalist Canadian Ben Johnson, running a scarcely believable 9.79 s in the Seoul Olympics 100 m sprint. Growing up in Britain, and studying biochemistry in Canada, my youth was full of the pain of Sharon and the shame of Ben.

The phrase 'drugs in sport' comes with extensive emotional cultural baggage. When people think of drugs in society at large, their views are coloured by the roles played in medicine and recreation. A 'drug' is a powerful external agent. When prescribed by a doctor it can heal illness or induce painful side effects; when taken on the street it can leave you savaged, addicted and ruined for life. So we should not forget that the word 'drug' itself is evocative and we take that image into our views of the effects in sport. Using the word *drug* gives the public an excuse to stigmatise something they find unacceptable. But it also benefits the drug users. Appropriate labelling enhances the benefits of their performance aids via the powerful placebo effect—for who wouldn't run faster with a magic custom *drug* in their body? Even Willy Voet, in his confessional book about doping in the Tour de France, admits that sometimes giving just a simple sugar solution had amazing results.[2] All he had to do was tell his cyclist it was a drug that 'could be anything and everything'.

Names also matter. Pharmaceutical companies agonise over what to call a drug. They know that a good name will improve efficacy as well as marketability. Names are targeted to treatments; the same compound (sildenafil) is called Revatio when used to treat arterial hypotension and Viagra when used for erectile dysfunction.

Given this backdrop, it is notable that the main agency testing for performance-enhancement, the World Anti-Doping Agency (WADA), never uses the word 'drug'—instead preferring 'doping' and 'doping offence'. However, in popular sporting culture there is no way to avoid this nomenclature, nor the effect it has on the athlete's environment. The link between society's views of recreational drugs and its view of sport doping is clear—not least because many recreational drugs incur a doping offence, even when it is very unclear what sporting benefit they can provide.

A history lesson

There are many parallels between recreational and performance-enhancing drugs. Both have only recently become unacceptable. For although drug abuse in society tends to be considered a modern problem, it is really the negative response to drug use that is the new issue. Opium was the recreational drug of choice in the Ottoman Empire and in the New World coca was considered so innocuous that it featured in the original version of the most famous of soft drinks that bears its name. Indeed the US government even took the Coca Cola Company to court, accusing it of not including enough coca in its product to justify use of the brand name and, even worse, replacing it with caffeine, which was considered at the time to be a more dangerous drug. This led to the wonderfully entitled 1911 court case 'The United States v. Forty Barrels and Twenty Kegs of Coca-Cola'.[3]

A similar lack of concern about chemical enhancement holds in the field of sporting endeavour. We can go as far back as the ancient Egyptians who allegedly used boiled hooves of asses to improve their performance. Indeed the possible benefits of different ergogenic aids seemed to be as hotly debated then as today. Charmis of Sparta swore that dried figs led him to Olympic gold in 668 BC, whereas one of the fathers of modern medicine, the second century Greek physician Claudius Galen, noted the positive benefits of eating herbs, mushrooms and testicles. Galen was the physician to the gladiators, the kind of 'sport' where you can't get away with a bad day at the office and optimal preparation of body really matters.

Debates continued to rage about diets into the third century. In the view of Philistratos (AD 200) the Ancient Greeks 'made war training for sport and sport training for war.'[4] But he bemoaned his generation of sportsmen who 'spent too much time eating, drinking and fornicating instead of actually training' and who treated sports 'more of a hobby than a way of life'. This was reflected in their poor choice of nutritional aids. Greek athletes of his generation ate white bread, poppy seeds, fish, and pork while the ancient Spartan athletes trained on a meat-full diet of bulls, oxen, goats, and deer.[4] Perhaps too much should not be made of the details of Philistratos' complaints. After all he does sound all too much like a grumpy old man evoking the splendours of yesteryear. Indeed modern studies suggest that Galen's gladiators ate a quite un-Spartan diet of barley and beans washed down with a vinegar/ash 'sports' drink.[5] And as for fornication, in AD 77 the famous Roman author Pliny the Elder suggested that sluggish athletes were actually revitalised by sex.[6]

Philistratos and others might have bemoaned the nature of the ergogenic aids used in their day, but it is difficult to find any notion in the ancient world that using a particular compound or diet was unfair or should be banned. The debate, as far as we can judge, was about methods not morals. The view seemed to be that any way to obtain an edge was fine. Dietary supplements and extracts were just another way to achieve this. The use of performance enhancing chemicals continued from the ancient to modern era unabated and unchallenged. In 1904 Thomas Hicks won the Olympic Marathon in St. Louis on a combination of strychnine injections laced with brandy. This didn't seem to concern the authorities, although maybe they were distracted by chasing the original winner, Fred Lorz, who was disqualified for covering eleven out of the twenty-six-mile race in a car.

In the inter-war years, German and American scientists and coaches were openly experimenting with a wide variety of performance-enhancing compounds, including cocaine, adrenal hormones and amphetamine. Britain was not immune. Arsenal football club happily handed out stimulant pep pills before a key 1925 cup match against their local rivals, West Ham United. They were told these pills would enable their players to 'put in shots that looked like leather thunderbolts'.[7] The West Ham keeper was untroubled: the match ended in a 0–0 draw.

Although the Arsenal players and manager were happy to discuss this drug use in a very matter-of-fact manner in their biographies, at the time the football club did have some qualms about openly advertising their methods. No such false modesty accompanied rival football club Wolverhampton Wanderers a decade later in the 1930s. They even informed the media of their latest pharmaceutical tricks, publicising their use of extracts of monkey glands in the newspaper the *News of the World*. Although some people were concerned about potential health risks, there seems to be no great public outrage about 'cheating' and the use quickly caught on, with both FA cup finalists stoking up with the glands before the 1939 final.

However, the late 1930s did see the first signs of the sporting bodies beginning to act. The amateur International Olympic Committee banned the use of 'drugs or artificial stimulants of any kind' in 1938. As ever professional sport moved slower; there was speculation that a 1939 meeting of the Football League would address the increasing use of gland extracts but this came to nothing. More significant world events were about to intervene to disrupt sport. Yet World War II also had its pharmaceutical angle; there was extensive use of amphetamines by the armed forces of both sides.

In the 1940s amphetamine was in widespread use in society and not just for medical use. Society at large, whilst being warned off heroin and cannabis, was happily taking amphetamines as Benzedrine inhalers or in tablet form. People were using it to enhance mood, increase alertness and decrease the need for sleep. Amphetamine use in society was mirrored in literature. Whilst not surprisingly 'bennies' feature extensively in the novels of the beat generation guru Jack Kerouac, those of us who grew up on the James Bond movies will perhaps be surprised that the gadgets that aided him in his novels were as much biochemical as electronic. Forget a vodka martini 'shaken, not stirred'—in the book *Moonraker*, a benny slipped into his champagne enables Bond to stay alert for an all night card game. This doping scene was mysteriously missing from the movie. Not surprisingly sport mirrored society, with widespread use of drugs, mostly stimulants throughout the 1940s, '50s and '60s.

As in society, so in sport, drug use was increasingly being considered negatively by the 1960s. This led to a drugs crisis with the governing powers making serious attempts to ban the 'evil' of drugs in sport and society. In the USA, Benzedrine inhalers were made illegal in 1965, with the hallucinogen LSD (lysergic acid diethylamide) following in 1968. In sport, drug testing was finally introduced in 1965 to back up bans on stimulant use. There was open resistance to this clampdown by the individuals concerned. At the same time as Californian hippies were protesting the criminalisation of LSD, Olympic cyclists were boycotting amphetamine drug tests.

Yet by the 1970s drug use in both sport and society had become publicly unacceptable. In 1971, President Nixon declared, 'America's Public Enemy No. 1 is drug abuse'. In the sporting world this period coincided with the increasing use of anabolic steroids, rather than amphetamines as the drug of choice to enhance performance. Unlike amphetamines, steroids had never been in general use by the public. There was no positive narrative that athletes could access to gain public approval. There would be no steroid equivalent of Tommy Simpson, the amphetamine-fuelled British cyclist who has a hero's memorial to his 1967 death on the slopes of Mont Ventoux in France. Today's narrative between drug 'cheats' and anti-drug 'heroes' has became firmly established in both the sporting and recreational arenas.

Pharmaceutical developments

The parallels between drugs used in sport and society relate to more than just cultural associations. They both share a common source for the products they use. The pharmaceutical industry has been the major driving force for all drug use in society. Indeed most recreational drugs were accidental by-products of attempts to create medically useful products. Heroin and cocaine were originally used as very effective pain killers, LSD was a by-product of attempts to make new central nervous stimulants and MDMA (3,4-Methylenedioxymethamphetamine, widely known as 'ecstasy') was a failed attempt to produce a product to prevent abnormal bleeding. Likewise with drugs used in sport: anabolic steroids were used to treat muscle wasting, erythropoietin (EPO) to treat anaemia, and human growth hormone for childhood growth deficiencies.

Not content with using existing products, people eventually began bypassing the medical market to develop new 'designer' drugs solely for recreational use. This began in the 1960s and 1970s with the development of minor chemical modifications to illegal hallucinogens such as LSD and PCP (Phencyclidine, also known as 'angel dust') in order to avoid criminal prosecution. It took sport longer to catch up, but in the late 1990s US amateur chemist Patrick Arnold started to synthesise and distribute drugs that had never been used in patients. These included norbolethone—a steroid abandoned by Wyeth Pharmaceuticals and never marketed for human use—and tetrahydrogestrinone (THG), the world's first completely novel designer anabolic steroid for sporting use. As with the manufacturer of designer LSDs, the main rationale was not to produce a more effective molecule but to avoid punishment by the authorities. In the case of the recreational designer drugs the intention was to design a molecule that even if it were discovered would not be covered by legislation. In the case of sporting designer drugs, the idea was to avoid detection in the first place by being immune to doping controls—hence the nickname of THG as 'the clear'.

For these designer drugs, the authorities were ultimately victorious. Tim Scully and Nicholas Sand were prosecuted for making the acetyl amide of LSD, known as ALD-52. They were convicted on the basis that, although ALD-52 itself was not illegal, they had to have possessed illegal LSD as a starting material for its chemical synthesis. Patrick Arnold had his come-uppance when a disgruntled co-conspirator, coach Trevor Graham, anonymously mailed a syringe of THG to the authorities. A test was developed and caught a number of athletes, most notably the British sprinter Dwain Chambers. This led to the BALCO (Bay Area Laboratory Co-operative) scandal and the resulting US Grand Jury subpoenas of a number of famous athletes.

What about the future? Can the dopers break free of the intellectual shackles of the pharmaceutical industry and develop genuinely new biochemicals that surpass the current limits of human performance? Will novel molecules be developed in private laboratories run by rich mad scientists, resulting in a *Jurassic Park* future of genetically engineered and chemically super-enhanced sportspeople? It is easy to get carried away with these questions. But in many cases the fantasy is without foundation. The internet has indeed brought a community together that are happy to sample any research chemical they can get their hands on for possible psychoactive or performance-enhancing effects. However, the development of genuinely new molecules is still likely to require the budgets and intellectual effort that only the big pharmaceutical companies, or the worldwide scientific community can muster.

Even so-called new recreational designer drugs such as desoxypipradrol, methylenedioxypyrovalerone (MDPV) and methylmethaqualone (MMQ) are based on established drugs, published patents or are obvious derivatives of these. And if this is so for recreational drugs, where serious profits exist, it is doubly true for the sports performance market. The fame may be large in sports, but the profits for any new compound—at least compared to the pharmaceutical and recreational drugs market—are likely to be relatively small.

What this means is that knowledge of the science of the future is already in our hands. The future ideas in performance-enhancing drugs are unlikely to come from secret laboratories funded by rich benefactors intent on creating a new species of elite athlete. The ideas are out there already. We can look at the scientific evidence in human and animal studies to see what works now and what could work in the future. We can confidently explore the range of biochemicals that have the ability to enhance human performance by the traditional means of reading the scientific literature and applying our knowledge of human performance. We already have access to the knowledge to understand what has been done, what is being done and what will be done to improve sporting performance. That indeed is the purpose of this book.

How do we know that drugs work?

To answer this question we will need to understand the molecular mechanisms underpinning the performance-enhancing chemicals currently on the market. What are these chemicals? Athletes are, not surprisingly, secretive about their doping regimes. Can we ever be sure exactly what an athlete is taking? We do have some useful pointers. Detailed notes from the East German doping regime of the 1970s are now available and the recent US court cases over the BALCO scandal (vide Dwain Chambers) give real insight into the doping methods and rationales used by coaches. Doping was not restricted to simple anabolic steroids in East Germany [A] or the UK [B]

> [A] 'In this drug administration program > 2000 athletes preparing for international competitions were treated each year. In addition, numerous "cadre B and C" and junior athletes,

including minors, were also treated with androgenic hormones and with substances such as human chorionic gonadotropin (hCG) and clomiphen, which stimulate endogenous testosterone synthesis'.[8]

[B] 'Dear Dwain,

Per your request, this letter is to confirm I am willing to assist you in providing UK Sport and others with information that will help them to improve the effectiveness of their anti-doping programs. The specific details regarding how you were able to circumvent the British and IAAF anti-doping tests for an extended period of time are provided below. Your performance enhancing drug program included the following seven prohibited substances: THG, testosterone/epitestosterone cream, EPO (Procrit), HGH (Serostim), insulin (Humalog), modafinil (Provigil) and liothryonine, which is a synthetic form of the T3 thyroid hormone (Cytomel).

Yours sincerely,
Victor Conte'[9]

So we do have some insight into the kind of methods coaches are actually using in elite athletes. This is a very useful starting point. But how do we know something is performance enhancing? This is not as simple a question as might first appear. Just as with medical drugs, so in the world of sports doping, early promise in the laboratory is rarely matched by an equivalent effect in the field. In both cases part of the problem is the nature of the subject. In the case of medicine a drug is trialled on healthy animal models and people before being tested on an abnormal and heterogeneous subject group—unhealthy patients. The multitude of interacting pathologies (and even simple old age) can reduce the effect when the drug is finally marketed.

The same effect occurs in sport, but for almost completely the opposite reason. The test subjects are not healthy enough. Laboratory tests are almost never on elite athletes, as volunteering for a legitimate study would disqualify them from competition. Instead normal healthy athletes are used. However, elite athletes are abnormal. We are not even sure that the biochemical mechanisms underpinning a performance enhancement in the average athlete are exactly the same as in the elite athlete. It is entirely possible that by a combination of genetic predisposition and extensive training, the elite athlete will have already acquired the improved performance a drug or ergogenic aid purports to yield. Even though the difference between winning and losing is often measured in fractions of a per cent, laboratory success does not always translate to performance benefit. With the possible exception of the East German STASI (Ministry for State Security) documents, we do not have scientific tests on elite athletes following the administration of potentially performance-enhancing drugs.

Some useful performance tests of individual compounds have taken place in controlled laboratory settings. Yet in the real world the presence of other drugs and nutritional aids can interfere with the drug being tested. Athletes are notoriously secretive even about their legal training methods and ergogenic aids, let alone the doping agents they are using. They use drug regimes that are impossible to reproduce ethically in laboratory or animal studies. For example to study scientifically the cocktail of drugs Dwain Chambers used we would need to dope a population of fit athletes, remove one compound at a time from the list and explore the resulting performance effect. This is far too complex and time consuming, even if it were ethical. Patients are frequently excluded from testing new treatments in clinical trials if they are on additional medication as it can confuse the analysis of the data. Any patient with the drug combination Dwain Chambers was using would not have the remotest chance of being selected to be a subject for a scientific study.

For all these reasons a ten per cent increase in performance in laboratory trials on 'normal' subjects never results in a ten per cent increase in elite performance in a sports event. Nevertheless we are stuck with this subject group. There is a fine quotation about the scientific method in Robert Pirsig's philosophy book *Zen and the Art of Motorcycle Maintenance*.

> When you've hit a really tough one, tried everything, racked your brain and nothing works, and you know that this time Nature has really decided to be difficult, you say, 'Okay, Nature, that's the end of the nice guy,' and you crank up the formal scientific method. There's no fault isolation problem in motorcycle maintenance that can stand up to it.[10]

When testing the efficacy of any treatment in humans, whether medical or sporting, the randomised double-blind placebo-controlled trial is the equivalent of Pirsig's 'formal scientific method'. Uncreative, boring and tedious to implement, but no one can argue with the results. However, unlike other popular science books that dissect the effect of pills and medical treatments, you will now be aware that we cannot employ this tool to full effect when studying aids to top performance in human athletes. The truth is out there, but we will need to use our detection skills and scientific intuition as much as brute force clinical trials.

Can evolution be beaten?

What are the limits of human performance? What are the areas where ergogenic aids and drugs really could help? In some ways these are questions about evolution. If human evolution has optimised performance in certain activities, it may be more difficult to enhance them artificially. But if a particular sport requires skills that do not relate to something that we have evolved to do, there may well be scope for massive pharmaceutical improvement. This argument can be likened to a comparison of the benefits of sunscreen. Someone evolved in a hot, bright climate with suitably dark skin has little need of chemical enhancement to survive in the midday sun. On the other hand someone with light skin will burn. Only in the latter case will the chemical enhancement (sunblock) be of benefit.

So what kind of physical performance have we as humans evolved to optimise? As ever when looking into the evolutionary past there is uncertainty. It has been suggested we are poor runners compared with many other mammals, as the evolution of our brains enabled us to use tools rather than our bodies to hunt. We have lost our tails, an essential component of fast runners such as kangaroos and cheetahs. When it comes to sprinting, we won't win many awards in the animal kingdom. So maybe running is a largely irrelevant adaptation, piggy-backing on our need to walk upright? If natural selection has not acted primarily on speed or endurance running there could be a lot of capacity to improve on evolution. Could drugs act where evolution hasn't?

At least when it comes to endurance running, recent anatomical and anthropological studies suggest that this argument may not be strong.[11] We have several adaptations for running; these include long, springy legs and big, muscular buttocks that, at least in part, can replace our lack of tails. Our relatively long neck allows our shoulders to twist independently of our head as we gaze forward. But more importantly we lack hair, sweat a lot and have a very thin skin with blood vessels close to the surface. All of these adaptations maximise heat transfer from the body to the environment. Why does this matter? Due to the massive increase in oxygen consumption, animals can generate as much as six times the heat when running compared with lying down. This puts a major break on long distance running—you can literally burn yourself out if you don't stop. In one, possibly apocryphal experiment at Harvard, biologists put a rectal thermometer in a cheetah and found that once it hit 105°F (41°C) it stopped, even though it was running well below its normal fastest speed. It seems this is less of a problem for humans. We have evolved, at least in part, for slow long running. This enables us to track our prey by a technique called 'persistent hunting'. This strategy involves targeting one animal and tracking it exclusively, usually starting in the hottest part of the day. As long as the same animal is hunted we will generally win our meal—even if it has to take over a day.

While not exactly having the tracking skills of today's persistent hunters—the Bushmen of the Kalahari or the Tarahumara of northern Mexico—I have on occasions revelled in my superiority as a hunter. Admittedly my prey was just my garden rabbit who was avoiding being returned to her cage. After wasting many hot summer days unable to outrun her, I changed strategy to walking after her slowly and persistently. Eventually, after about ten minutes brisk walking, I could easily catch my exhausted prey. The same skills can be seen when humans race animals competitively. When running for a whole day humans have the capacity to outperform other species that are far faster over shorter distances such as dogs and horses. Twenty miles seems to be the tipping distance; the village of Llanwrtyd Wells in Wales hosts the annual man versus horse marathon. Usually, but not always, the horse wins. The only time men won both first and second place was on a very hot day, perhaps proving that as a species we run better in the heat.

How about strength? Are we humans optimised for power events? Again the evolutionary history is not straightforward, but most commentators suggest the evolutionary path that led to *Homo sapiens* involved a decrease in power capacity. Certainly all the other current great apes are much more powerful and muscular than the average human. And while it is not easy to determine power from skeletons alone, rival species like Neanderthal man certainly seem more likely to have been stronger than the co-existing modern humans. Gary Sawyer, an anthropologist at the American Natural History Museum in New York[12], says, 'They had very strong hands. If you shook hands with one, he would turn your hand to pulp.' Theories of the extinction of Neanderthal man 27,000 years ago have included suggestions of direct attacks by our ancestors. Yet no one is suggesting the success of *Homo sapiens* was due to raw physical attributes.

Conclusions about modern attributes of man based on evolutionary arguments always have the danger of sounding a bit like a 'Just So' story. But at the risk of making a dangerous prediction, I will venture to say that my reading of the evolutionary runes is this: if there is any sporting endeavour where drugs and ergogenic aids are *least* likely to have an impact it is in the field of the ultramarathon. 'Slow and steady' has already won the evolutionary race.

The case for sex

There is one glaring exception to all these ideas. Evolution has not acted equally on men and women. And this is most clear when it comes to the allocation of genes that control power and strength. The reason for this is not as obvious as it might seem. It is unlikely to be a product of differential hunting requirements. Indeed in many mammals it is the females that do the hunting. There seems no reason, apart from cultural, why a woman could not hunt as effectively as a man, whether over a short period or persistently for two days—and of course great apes that only hunt vegetables (gorillas and orangutans) still have males that are considerably more powerful than females. Instead the selection for strength is most likely a product of males fighting other males in competition for mates.

The evolution of human sexual dimorphism (the differences between the sexes) is highly contentious. Witness the controversy surrounding E.O. Wilson's book launching the subject of Sociobiology and the resultant heated debates about the evolutionary underpinning of human behaviour.[13] Perhaps this is not surprising as the areas under discussion include sexual preference, sexual behaviour and intelligence. With regards to less contentious issues such as physical strength you might think there was more consensus. But surprising traps exist for the unwary even in this area. The earliest bipedal hominid was the extinct species *Australopithecus*. You would think that there would be an accepted wisdom about basic structural dimorphism (at least with regards to the size of the male and female skeleton). But a debate still rages over claims that our earliest two-legged ancestor had males as much as fifty per cent larger than females.

Whatever the exact details, it seems likely that males and female hominids have indeed become similar in size with the passage of evolutionary time. But they are not there yet. Except in those rare events where strength is not a factor at all, we still have distinct male and female sporting categories. What this means is that evolution has allowed a large scope for improvement to any female who can increase their male physical characteristics.

Victor Conte founded the infamous Bay Area Laboratory Co-Operative (BALCO) that was accused by the US Anti Doping Agency of supplying anabolic steroids to a wide variety of athletes including Dwain Chambers; in 2005 Conte pleaded guilty in the US court to conspiracy to distribute steroids. This is what Conte said about doping female athletes,[14] 'Steroids can help a female sprinter to lower her 100 m time by about four-tenths of a second or four metres faster. The effects of steroids upon male 100 m sprinters are about two-tenths of a second or two metres faster.'

Whilst not necessarily agreeing with this statement quantitatively, qualitatively it is sound. Female world and Olympic records set prior to random drug testing have been much harder to break. For example, whilst there is a steady progression in the male Olympic athletic records, there are as many female Olympic records still standing that were set prior to 1990 as those that were set in the last decade (see Figure 3). It is hard not to argue with the implication that the steroid doping that was widespread in the 1980s has had a more dramatic effect in female sport than male sport.

But the male/female issue has other implications. There are methods other than steroids to close the gulf in performance that exists between male and female sports. The simplest is by direct subterfuge. The 1938

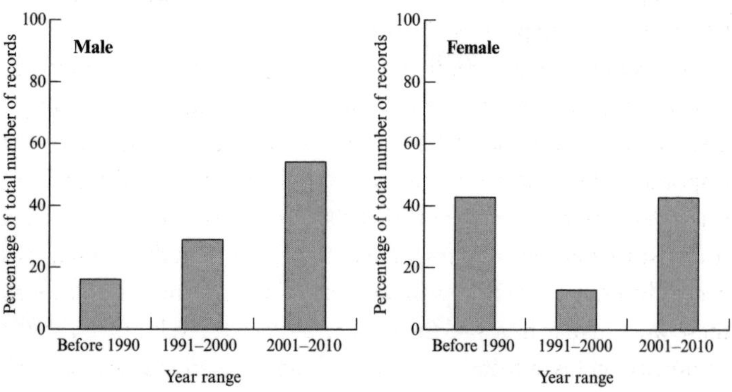

Figure 3 A comparison of male and female Olympic records in athletics

The graphs indicate the year range in which a current Olympic record (as of 2008) was set.

European Champion high jumper, Dora Ratgen, is perhaps the most famous example. Ratgen's case featured in the film *Berlin* 36. In it she was unmasked as a transvestite member of the Hitler Youth picked for Hitler's 1936 Berlin Olympics in place of the genuinely female Jewish athlete Gretel Bergmann. However, on closer inspection[15] this case is more complex. There is no evidence that there was a Nazi conspiracy to infiltrate a transvestite into the team. In fact Dora Ratgen had somewhat ambiguous genitalia at birth, was registered as a woman and brought up as a girl. The Nazis were unaware that she was a man. Given the nature of the times it is unlikely the German authorities required such an elaborate subterfuge to discriminate against a Jewish athlete anyway. Upon being discovered in 1938 and being shown to indeed be genetically and physiologically male, Dora quit sport and changed her name to Heinrich.

Such direct subterfuge was very rare in the past and is probably non-existent now. Yet there are more subtle grey areas between what constitutes a male and female athlete. Suspicion about the gender of supposedly female athletes led the International Olympic Committee to introduce gender tests at the 1968 Olympics. These degrading and simple physical examinations were later replaced with a scientific chromosome test for the presence of the male Y chromosome. But even this is not foolproof. At the 1996 Olympics in Atlanta, 8 of 3,387 female athletes were found to possess a Y chromosome.[16]

How could a person have female characteristics, despite the presence of a Y chromosome? In seven of the cases, the cause was Androgen Insensitivity Syndrome (AIS), a condition where the body does not respond to the male hormone testosterone; in one case the problem was a deficiency in the enzyme that activates testosterone. The lack of response to testosterone during foetal development in these conditions generally results in external genitalia that are mostly female. So although these women may have had the same testosterone levels as men, it was assumed that this testosterone could not lead to a performance enhancement and the athletes were allowed to compete. However, whilst the prevalence of AIS in the general population is 0.002 per cent, in women competing in the Olympic Games it is 0.2 per cent. This 100-fold increase, suggests the possibility of some performance enhancement due to this syndrome.

Most people with AIS syndrome have XY sex chromosomes, but still consider themselves female. The International Olympic Committee now think the same—they abandoned compulsory gender testing after Atlanta. But they retained the right to test for gender if necessary and the suspicion existed that if a genuine world-beater emerged with ambiguous sexuality the story might be different. And this was precisely what happened with the recent cases of Santhi Soundarajan and Castor Semenya. Soundarajan failed a gender test at the Asian games in 2006 after winning the silver medal in the 800 m race. Castor Semenya was a convincing winner in the Berlin 2009 World Championships in the same event. She was subsequently withdrawn from international competition pending an investigation. Eleven months later the International Association of Athletics Federation cleared her to compete. It has been suggested that as part of this process she is required to undertake therapy to reduce her testosterone levels.[17]

Is it all in our genes?

The clear male/female differences in sport performance do raise interesting questions about athletes who are genetically rather than pharmacologically enhanced. If a genetic mutation that makes a female athlete more 'male' makes that competitor ineligible, what about another that makes a male athlete run faster or jump higher? After all, elite athletes are all genetic anomalies—it is just a question of how extreme the anomaly. Will there soon be DNA testing to accompany drug testing?

Sport has so far avoided this problem by assuming that performance is multifactorial. To be an elite athlete, it is wise to choose the right parents, but no one has previously thought that this is due to inheritance of a single gene modification. Unlike in TV shows, you cannot inherit superhero abilities. Yet recent research has raised serious doubts about this cosy hypothesis. In many domestic and laboratory animals it is indeed true that a single gene can make a dramatic difference in strength and endurance. And there is a precedent in elite sport.

Eero Mantyranta, a Finnish cross-country skier who won gold medals in the 1960 and 1964 Winter Olympics was found to have abnormally high levels of the protein haemoglobin in his red blood cells. High haemoglobin levels allow athletes to carry more oxygen and are a benefit in endurance sport. Mantyranta, like all his family who were tested,[18] achieved their high haemoglobin levels by having a genetic mutation in the protein in their body that responded to the hormone erythropoetin (EPO). At normal levels of EPO the effect on their body was as if they had much higher levels. A stronger EPO effect means more haemoglobin and therefore more oxygen to their cells. EPO is top of the list of genuinely performance-enhancing chemicals; yet here is someone who through an accident of birth behaves exactly as if he was doping with EPO. Is this fair? How is it different from the case of Santhi Soundarajan and Castor Semenya? The more we know about the genetic make-up of elite athletes the harder our decisions about what is right and proper in sport will appear. In ten years' time full genetic profiling may be commonplace. We may also have knowledge of some rather specific genetic adaptations that improve sporting performance. What do we do to ensure fairness in competition in this case? Doping control will not be enough. The Paralympics currently has a wide range of classes that reflect the differing physical abilities of athletes. Will the same be true of the rest of the Olympics? Will there be different classes dependent on different genetic sequences in key molecules in the body? I once would have thought this a fantasy. Now I am not so sure.

I will explore the intricacies and complexity of artificial genetic manipulation in the final chapters of this book. But for now there is enough to explore in the kind of drugs used in the two races described in the Prologue. Sporting achievement requires power, endurance and the ability to compete beyond the pain barrier. The unholy trinity of anabolic steroids, EPO and stimulants can impact in these areas. But before we study these drugs there is a need to explore in more detail the factors that can limit athletic performance.

The Limits Of Human Performance

'No one can say, "You must not run faster than this, or jump higher than that." The human spirit is indomitable.'
Sir Roger Bannister

Is it really so indomitable? There must ultimately be a limit to human sporting performance. The ability for an athlete to perform at their optimum is dependent on three factors, the athlete's body, the task to be performed and the external environment of the competition. Although these factors are intimately connected, the use of appropriate chemicals might be expected to affect predominantly the first of these factors (the human). The problem is it is not always easy to determine which factor is acting in which sport at which time.

In the introductory chapter I implied that sudden changes in sporting performance were suspicious and could be attributable to athletes discovering a new doping method (performance increase) or the fear of a detection resulting in a fall in doping use (performance decrease).

However, especially in complex sports there are frequently alternative explanations for sudden performance increases (or indeed decreases). As often or not these relate to factors external to the athlete's body.

The physical environment—what's in the air?

Anyone who has thrown a Frisbee in the garden or flown a kite in the park knows that the wind has a dramatic effect on the performance of objects. These objects include human beings, which is why many athletic world records are invalidated if there is too strong a tail wind. In 1988 the fastest woman in the world, Florence Griffith-Joyner, ran 100 m in 10.49 s, a massive 0.15 s faster than any time that has been run since. 'Flo-Jo' ran very, very fast for one season and retired immediately, leading to speculation (not proven) that she achieved her results with pharmacological aids. Recent studies suggest that, at least in part, human error was to blame. The wind speed for races before and after she ran was measured at over five metres per second behind the sprinters, much greater than the two metres per second limit allowed. However, when Flo-Jo ran the 100 m in 10.49 s the wind speed gauge returned a mysterious 0.0 reading, allowing her to record her sensational world record time. A subsequent analysis has suggested that it is likely that the device was not plugged in at the time the race was being run.[1] Sometimes athletic performances really are too good to be true.

Actually while having the wind behind you is a good thing, it does not compensate for the penalty of having to struggle through a head wind. Athletes certainly feel when running on an oval track that they lose more time in the headwind than they gain with a tailwind.[2] Interestingly the scientific reason for this is not what you might expect. Athletes running into a very strong wind change their running style and convert drag into body lift, some even reporting that they almost 'fly' between strides. In contrast, running with a strong tail wind is mechanically inefficient. So although running with the wind behind you increases your speed somewhat, the benefit is not sufficient to overcome what is lost in the head wind, resulting in an overall time increase for a windy lap.

The amount of air (or air pressure)—as well as its speed—has a significant biomechanical effect on human performance. The less air you have, the lower the pressure and the lower the air resistance. One way to reduce air pressure is to climb a mountain. The biochemical cost in terms of energy expenditure is always less at high altitude. But less air also means less oxygen. In the Mexico City Olympics in 1968, held at 2,240 metres above sea level, the short-distance track races were run better than before with many world records; times suffered for the long distance races where oxygen was more important. The cross-over point for men where there was neither benefit nor loss was at 800 m (twice around the track). Interestingly for women, 400 m seemed to be the limit of benefit.

Any event that involved throwing an object—whether physical or biological—likewise benefited from the lower air pressure in Mexico. One of my earliest sporting memories is that of the American Bob Beamon's long jump. Shattering the previous record by almost two feet, he carried on jumping and bouncing when he hit the sand pit, eventually ending up on the track itself. 'Compared to this jump we are as children' was the quote of the previous world record holder, the Soviet jumper Igor Ter-Ovanesyan. This is the reason why world records in these types of events are appended with '(A)' if achieved at altitude and why footballs fly further through the air in Denver's famous 'Mile High Stadium'.

The force of gravity brings all things down to earth in the end. As astronaut Alan Shepherd showed it is easy to hit a golf ball further than Tiger Woods if you are on the moon. Shepherd had the benefit of both low gravity and no air resistance, though I guess was somewhat encumbered by having to drive the ball whilst in a space suit. But what are gravity effects at the Mexico City and Denver stadia? In fact the effective gravitational force changes relatively little due to the altitude above sea level; it is far more dependent on the distance from the equator. Gravity (and hence your weight) increases by about 0.5 per cent at the poles compared to the equator; even climbing Mount Everest only yields a

0.3 per cent reduction. A person would weigh more if they lived in Scotland rather than England and would lose weight the further they moved south of the border—although you would need a very sensitive pair of scales to measure this.

What about sporting performance? Mexico City's gravity is about 0.2 per cent less than Denver's, most of this being due to its proximity to the equator rather than the relatively small increase in altitude. What effect do these gravitational changes have on performance? It is difficult to calculate precisely, but any effect on Bob Beamon's long jump record is likely to be small (maybe a centimetre or two). For longer distances a change in the gravitational force could be significant though. At the time of writing (2011) the British jumper Jonathan Edwards holds the current world triple jump record for a jump in Gothenburg, one of the world's 'heaviest' major cities. His 18.29 m hop, step and jump could maybe have been as much as 5 cm further if he had not been weighed down by the force of Swedish gravity.

How could chemical enhancement affect an athlete's ability to interact with gravity, air pressure or wind resistance? One option is to lose body mass. This has an obvious beneficial effect on overcoming the force of gravity, as long as it is not at the expense of a drop in strength. There are several drugs that cause weight loss, for example the banned diuretics that lower body water content. However, no magic gravity-defying drugs exist as yet. On the other hand there are two properties of air that limit human performance and which can readily be affected by chemical enhancement. These are the temperature and the oxygen content.

Whilst in evolutionary terms humans have readily adapted to hot climates, adaptation to prolonged cold exposure has proved much more difficult, requiring significant behavioural adaptations. The micro-environment in an Inuit igloo for example is maintained at a balmy 21°C; outside activities are performed in clothes designed to mimic the environment of a much warmer climate. In terms of acute exercise performance however, it is heat that is the enemy. Dehydration is the main problem, especially if high humidity impedes cooling by sweat evaporation. Even if optimal fluids are taken on board there is still a limit to the speed of water absorption by the gut. Worse, blood flow is redistributed away from the exercising muscle and towards the skin, enhancing the cooling process but reducing the efficiency of performance. Training in an appropriate climate can improve the ability to exercise in a hot, humid environment by increasing the water content of the blood and enhancing the efficiency of evaporative cooling from the sweat glands. But a performance hit is still taken. Cold weather is less of a problem. This is partly due to the fact that no activity that uses energy is 100 per cent efficient. Inefficiency results in heat production. Consequently exercise will always generate heat and therefore partly compensate for a cool environment. The consequence of this is that performance in long distance sporting events is always compromised by extremes of heat, but rarely of cold. Athletes take fluids to improve performance in hot weather. These contain more than just water; the use of the right ergogenic nutrient mix to enhance performance is critical. These may not all be entirely innocent. Recently drugs have become available that can prevent the debilitating effects of heat on performance.[3]

The amount of oxygen in the air we breathe is a clear limiting factor in performance. In extremis we see this when people try and climb Mount Everest without oxygen. The lower air pressure means that, even though the oxygen content remains at 21 per cent, the same amount of air has fewer oxygen molecules present. The body can adapt to this lower air pressure by altering its physiology. But performance is always compromised to some extent. Even at the relatively low Mile High Stadium in Denver, there is likely to be some effect on American football players who 'drop in' and 'drop out' for a game with no high altitude preparation. The effect is even more extreme in South America where the top football teams in Bolivia, Ecuador and Peru play at altitudes almost double that of Denver. In 2007 this led to the world football governing body, FIFA, banning competition at these heights unless players had time to acclimatise for one week (above 2,750 metres) or two weeks (above 3,000 metres). The ban was temporarily lifted, as some matches in the 2010 World Cup qualifying campaign had already been played. Not surprisingly the ban was favoured by the low-lying football powerhouses of Argentina and Brazil, and condemned by the Andean nations of Bolivia and Ecuador. Bolivia has the most to lose. It plays its home games at the capital La Paz (altitude 3,600 metres). Despite finishing next-to-last in the qualifying campaign, it still managed to win home games against top teams such as Brazil, Argentina and Paraguay.

Given that there is a clear performance decrease due to lack of oxygen at high altitude are there chemicals that can be used to prevent this? The only approved drug is Acetazolamide (sold under the trade name Diamox). When someone goes to high altitude, and the oxygen drops, their breathing rate increases to compensate. Acetazolamide enhances this increase. It thus can accelerate the process of acclimatisation, but it is not a magic bullet. People can still take weeks to reach peak performance when transferring to a high altitude. Despite the fact that sports discourage competition under these conditions there is still a strong drive to produce new pharmaceuticals to optimise high altitude performance. Wars fought in countries like Afghanistan involve fighting in both low-lying deserts and mountain ranges over 4,000 metres high. The increasing use of helicopters in modern warfare means that soldiers are rapidly transferred to these high altitudes. There is a drive to develop new chemicals and methods that can allow troops to avoid the lengthy need for altitude acclimatisation. The US Defense Advanced Research Projects Agency (DARPA) is the forward thinking, military funded, agency credited with the invention of the Internet. In 2010 DARPA spent $7 million per year on its 'combat effectiveness' programme, the main goal of which was to produce new products to alleviate the problems of high altitude.

When it comes to sport athletes make use of altitude training to enhance their performance at low altitude. Essentially the trigger of low oxygen in the air changes an athlete's biochemistry and physiology. It makes them able to use oxygen more efficiently and hence perform better at low altitude. This is an area which the dopers have exploited. The blood booster EPO works by short cutting much of the benefits of altitude training—predominantly the production of new red blood cells. In principle it should therefore aid altitude acclimatisation. But EPO thickens the blood and can have detrimental side effects. This risk appears too high for use by the military who shun this drug. However, the same risk appears not to be a deterrent for elite athletes.

The technological environment

In many sports the athlete's performance is intimately linked to human-inspired enhancements of their physical world. Whether it be an inanimate car or an animate horse many races depend on external objects; these can be at least as important, if not more so, than the athlete. Even in less obvious situations, the level of performance is affected by the environment. Ice skating stadia and athletic tracks can be prepared to optimise the chance of a world record. Indeed the running track used in the Mexico City Olympics was of a new artificial kind that could have contributed to some of the fast times that were initially attributed to the low air pressure.

Sometimes the technology becomes so enhanced that events have to be 'recalibrated'. In the 1990s tennis racquets made from new alloys enabled a greater power of serve. Fast serves that bounce at variable heights on grass tennis courts are difficult to return, leading to spectator-unfriendly short rallies. In 2001 the surface for the Wimbledon tennis championships was changed to a more durable 100 per cent rye grass; this resulted in a more regular bounce and consequently longer rallies. However, it has been suggested that this new grass surface had the effect of decreasing the performance of the British 'serve and volley' specialists at the time, Tim Henman and Greg Ruzedski—a rare example of a side removing a home field advantage.

More understandably changes sometimes have to be made for safety reasons. In the 1980s javelins were being thrown so far that they were in danger of leaving the athletics stadium and hitting spectators. The 1986 world record of 104.8 m was left unaltered and new records started with a newer, less aerodynamic implement with an altered centre of mass (the record now stands at 98.48 m). Manufacturers still try to make technological improvements. One particular example in 1991 resulted in a retrospective banning and rescinding of the resulting world record.

Equipment is not restricted to events requiring the athlete to move a physical object. The athlete's body itself is subject to the laws of biomechanics. Technology can have an effect here. Most notably in swimming, drag in the water is a major detriment to performance. Traditionally costumes were designed to be as small as possible with the rest of the body shaved (both males and females) to improve the hydrodynamics. There were advances in swimsuit design, but these tended to be gradual. But in 2008 a new textile swimsuit was designed that increased buoyancy, rather than just reducing drag. Swimmers could now swim higher in the water. Air is less of a drag than water and speeds dramatically increased. In summer 2009, twenty-eight world records fell at the World Swimming championships in Rome. This represented seventy per cent of all the records, an unprecedented amount at a single meet. These swimsuits have now been banned, but the world records remain.

A new phrase, 'technological doping', has entered the sports lexicon. The UK track cycling team that was so successful at the Beijing Olympics in 2008—winning seventy per cent of the gold medals on offer—had a bike development programme that certainly enhanced their speed on the track. The design of bicycles, suits, helmets or skates is as closely guarded by sportspeople as their training programmes themselves. It could be argued that the secrecy in this technology mirrors the secrecy in the shadier world of 'real' doping. But there are major differences. The technology is visible and testable at the moment it is being used. Consequently there is rarely an attempt to 'cheat'. Some of the notable exceptions seemed doomed to failure from the start. One is reminded of the example of Boris Onishchenko in the 1976 Olympics. Hits on an opponent in fencing are monitored by an electronic signal. Dubbed 'Disonischenko' by the world's media, the fencer wired up his épée to monitor a hit when he pressed a secret button, rather than actually needing to hit his opponent. Of course the effects soon became obvious to his opponents, resulting in the IOC disqualifying him from the competition.

Sometimes a technological improvement leads to a later enhancement that is not technology driven. The replacement of sand pits with deep foam matting enabled high jumpers to land on their backs with no concern for the fine safety details of the impact. This led to the introduction of head-first backward flopping styles. It is likely that this style was first used competitively by an American called Bruce Quande. But it took off when invented independently by more famous athletes with names more suited to the media's alliterative tendencies. Who could resist reporting on the 'Brill Bend' and the 'Fosbury Flop'?

One difference between technological doping and chemical doping is that the former does not have the potential to harm the health of the athlete. Indeed in that most technological of sports, Formula 1 motor racing, the advances in technology have resulted in safer cars. However, this is not a book on the technological limits of human performance. So is there a genuine interface between technological doping and the real thing? There is no doubt that part of excellence in sports performance is adapting to changes in the technology surrounding the sport itself. Every year a Formula 1 driver has to learn how to drive a new car for example. But can drugs help in this? It seems unlikely.

In the shot put, athletes throw a heavy metal ball as far as they can. Again a new style has influenced how far the ball can be thrown. Do you use the old-fashioned glide or the more elegant, modern spin? Both seem to have benefits. Short throwers are better at spinning around, whereas taller ones tend to glide. The body's biomechanics may alter, but in its essence this is a power event. Drugs can help with the power whatever the style adopted. Technology can modify the records in any sport. But it doesn't nullify the effect of drugs on the athlete's body itself. If a drug gives you more power, you will shot put further or swim faster no matter what swimsuit or throwing style you use.

The human environment

Most human activity does not take place in isolation. This is especially true for sport. It is well known that one of the key factors to control in any scientific laboratory test of sports performance is the extent of the encouragement, if any, given to the participant. The crowd, whether in a laboratory, a stadium or at home in front of their televisions, plays a key part in many sporting performances. There are some skill events where it is possible that performing in front of an audience is detrimental; indeed there are some athletes who perform badly under the fine microscope of public examination. However, in most cases where athletes have to perform a physical activity there is a performance boost when they perform the event in their home environment. Sometimes non-biological factors are at play like an unconscious home team bias by a referee or an away teams' unfamiliarity with a strange pitch. But home advantage is at its strongest when you have the home crowd behind you willing you on to win.

What is the scientific basis for this enhanced performance? Psychological tests indicate that people are more self-confident when playing in their home venue and are more anxious when playing in their opponents' venue. Is there a biochemical or physiological mechanism underpinning this crowd effect? Hormones control our biochemistry and we can measure them in saliva. Levels of the male sex hormone testosterone rise before a home event and increase more following success in front of a home crowd.[4] Sociobiologists explain this as a response to defending home territory from foreign invaders. Animals fight harder to defend their territory and their home, especially if they are protecting their young. Do we need to induce this kind of effect to optimise human performance? Can we mimic with drugs or ergogenic aids the effect of facing a carnivore? Is this why Floyd Landis took testosterone when he won cycling's Tour de France in 2006, resulting in his ban by the US Anti Doping Agency?

One problem is that in addition to the 'fight or flee' response to fear, humans also have a 'freeze' response, where their bodies stiffen and their heart rate slows down in an attempt not to be noticed. This is akin to a frog trying not to be noticed by a snake or, closer to home, a rabbit freezing in a car's headlights. We have all seen this response on a sports field, usually when our favourite team finally makes the Cup Final. Fight, flee or freeze it is likely that pharmaceuticals can play a role in allowing people to reach their psychological peak. So what exactly are the psychological limits to performance?

Psychological limits to performance

As in life, so in sport, the effect a role model has on performance is significant. In many cases once a significant milestone is broken, this leads to a rash of follow-up records performances. Perhaps the classic example of this is the sub-four-minute mile famously run by Roger Bannister in Oxford, England in 1954. Gunder Hägg's world record had stood at 4 min 1.3 s (4:01.3) for over nine years. But in the two years that followed Bannister's record breaking run, eight other athletes broke the four-minute barrier. Some of this is probably due to people copying details of the methods Bannister employed, including the use of fellow runners as pacemakers. However, it is likely that a lot was due to the psychological barrier. The words of Bannister's great rival, the Australian John Landy, perhaps sum this up best. Shortly after running 4:02 for the mile, Landy is quoted as saying:[5]

> Frankly, I think the **four-minute mile is beyond my capabilities**. Two seconds may not sound much, but to me it's like trying to break through a brick wall. Someone may achieve the four-minute mile the world is wanting so desperately, but I don't think I can.

Yet forty-seven days after Bannister's run, Landy broke Bannister's new world record with a time of 3:57.9, almost four seconds faster than he had previously run.

This effect of role models can be localised to individual countries. For example in my youth in the 1970s and 1980s, the world of athletics in Britain was dominated by the rivalry of two men, Sebastian Coe and Steve Ovett, who traded world records on what seemed a weekly basis in 800 m, 1500 m and the mile. But they rarely raced against each other. Even when they did, such as at the Moscow Olympics of 1980, they appeared inseparable. Ovett won the 800 m and, five days later, Coe strode past him to take the 1500 m. They were soon joined by other local heroes such as Steve Cram, Peter Elliott and Tom McKean. But now when I look at the world's best racers they are dominated by runners from Africa. While there has been a lot of discussion about a possible genetic component to this switch, the psychological effects of role models on performance should not be underestimated. For it is not just that Africans are running faster than British athletes; British athletes are getting slower. This can be seen clearly if we look at the fastest twenty times in the 800 m race. In world terms sixteen of these have been set post-1996. But if we just focus on the UK best times this number drops to just three out of twenty (see Figure 4). Shorn of positive role models the golden era of British middle distance running has imploded. The British now are just not as good as they were.

Similar effects can be seen in other countries. In the 1970s and 1980s the Caribbean islands of the West Indies had the most fearsome fast bowling attack in cricket. The likes of Andy Roberts, Michael Holding, Joel Garner, Colin Croft, Malcolm Marshall, Curtly Ambrose and Courtney Walsh regularly terrorised cricketers all over the world,

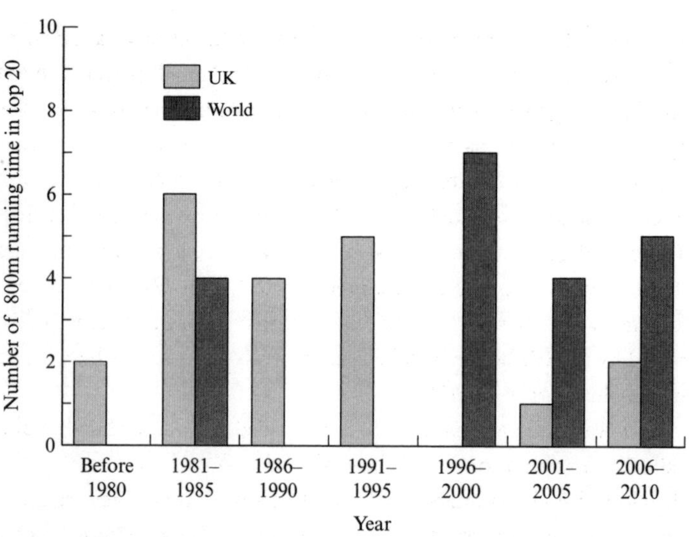

Figure 4 Distribution of fastest 800 m running times in UK and the World

leading to the introduction of the protective headware and body armour of today's cricketers. Yet today's Caribbean role models are not cricketers, but sprinters. For Michael Holding read Usain Bolt. The fall of West Indian cricket and the rise of its sprinters, past and present world-beaters from the same islands, says a lot about how culture can impact on sport performance. Even in today's global village we may be some way from knowing who can run, throw or jump the best if the person with these skills is not in the right place at the right time to express their activity.

Although the psychological effects on sports performance are complex, they are undeniable. With our increasing knowledge of the brain we are starting to drill down to the specifics of what stops us performing. What makes us stop running—the science of fatigue—is a very active research field at present. It is likely drugs can play a role in reducing this fatigue.

Shaping the human body: biomechanics

The shape of the human body and its musculature clearly has an effect on sports performance. High jumpers and basketball players are tall and flat-racing jockeys are small. Occasional extreme genetic anomalies appear that seem to have an unfair advantage. In many cases there are counter-balancing disadvantages. Most very tall people have difficulty in coordinating their bodies and take a consequent performance 'hit'. The old adage of a good big man will always beat a good little one doesn't always hold true. For example in 2010 Nikolay Valuev, the seven-foot tall world heavyweight boxing champion from Russia, was out-boxed by the UK boxer David Haye, who was eight stones lighter and nine inches shorter than Valuev.

However, size can matter if it comes allied with co-ordination and skill. There is a reason why basketball and volleyball players are taller than long distance runners on average. Perhaps a more interesting question is, does size matter when it comes to sports where a simple height/reach advantage is not quite so evident? This has recently been brought into focus by the world record sprinting of Jamaican Usain Bolt. In the 100 m race getting to the line first is a 'simple' function of stride length multiplied by stride rate. If a long-legged man can move his legs as fast as a short-legged man, then the greater stride length will win out in the end. In winning the Beijing Olympic 100 m, Bolt used 41 steps, compared to 44 steps of the other runners.[6] But getting the legs moving fast is a power issue. Usain Bolt has to be as muscular as smaller sprinters to deliver the power to get his long levers working at a similar rate. The key is in the co-ordination. Unlike Valuev's boxing, Bolt's sprinting is elegant and in control. We don't know how rare this ability is in a tall man. Will Bolt's success lead to a mass of tall sprinters who, unleashed from their psychological inhibitions, now realise they can run faster than the short guys?

We can get somewhere close to answering this question by looking at height statistics for a sport where athletes need to be both fast *and* tall. If the sport is popular and pays high wages there should be a strong selection for athletes with Bolt's physique (1.96 m height and 95 kg weight). A wide receiver in American football is required to run fast to get clear of the opposition defence, but still be tall and agile enough to catch a long pass from the quarter back. The average height of a wide receiver is 1.85 m with a weight of 91 kg giving a height/weight ratio of 0.020 m/kg. Bolt's height/weight ratio is almost identical at 0.021, but his height is six per cent above the wide receiver average. Therefore it is very rare to find a wide receiver as tall as Bolt. The only recent example was Ramses Barden, who played for the New York Giants. Barden is 1.98 m tall and weighs 103 kg. However, while clearly no sluggard, Barden is no Usain Bolt. In fact he was amongst the slowest of the 2009 crop of wide receivers.

We don't yet know whether Bolt is the first of a new breed of tall sprinters or a genetic anomaly we may not see the like of for very many years to come; but it is clear that body shape and the consequent changes in biomechanical efficiency can have dramatic effects on human performance. What role can ergogenic aids or drugs have on features that seem to be genetically fixed? Human growth hormone is a drug of choice for many athletes. However, the ends of our long bones fuse after puberty, fixing our ultimate height. Therefore whilst growth hormone is used to treat growth disorders in children and adults, only in the former case can height be increased. Its use by mature athletes is therefore based on its supposed short-term effects on strength, not any long-term increase in height. It is unknown what the effects of a superphysiological dose of human growth hormone would have on a human child. The systematic doping programme in East Germany, for example, did include pre-pubescent athletes, but growth hormone was not readily available at that time.

Growth hormone is made in the pituitary gland. Benign tumours of this gland usually cause pathological increases in the body's level of growth hormone. When this occurs pre-puberty it results in gigantism and this is almost certainly the underlying cause of all the world's very tallest people. Post-puberty excess growth hormone production results in a condition called acromegaly, whose symptoms include enlargement of the hands and feet, an expansion of the skull and protrusions of the brow and lower jaw. However, the structural changes associated with gigantism or acromegaly are, in general, not associated with improved athletic performance.

What about the opposite end of the scale? Are there situations where sports performance can be enhanced by artificially reducing height or weight? It is true that in some sports reducing weight in the short term is important for rather trivial reasons. A boxer needs to 'make the weight' to fight in his correct class and an overweight jockey adds an unnecessary load to a racehorse. In this case diuretics have been frequently used to reduce the fluid weight in the body. Even more controversially, some gymnasts have been accused of keeping their bodies artificially small, the rationale being that small bodies are more supple and able to be more effective in acrobatic and tumbling skills. As many top female gymnasts are very young, one way of keeping this small body is to prevent the growth spurts associated with puberty. In some gymnasts this occurs naturally via excess calorie restriction as they attempt to model themselves on the shape of current stars. While there have been documented cases of gymnasts banned due to diuretics, more sinister rumours persist that some athletes take drugs such as histrelin that can directly delay puberty.

Interestingly a very old-fashioned non-pharmacological method of creating smaller athletes has recently emerged in gymnastics—simply lying about your age. In 1981 a minimum age limit of 15 was introduced. However, as humans do not grow like trees adding a ring each year, it is impossible to accurately determine an athlete's age. There have been many cases of athletes and countries admitting to breaking this rule using false documentation. Sometimes the deceit is easy to detect. In 1993 the International Gymnastics Federation suspended the tiny North Korean athlete Kim Gwang Suk from the world championships in Birmingham (UK) when they discovered she had been claiming to be fifteen for three years in succession.

Visually the appearance of a body builder or an Olympic sprinter might appear very different to that of a 'normal' person. However, resistance (weight) training does not really alter the body's fundamental structure. Most, if not all, of the new muscle mass is due to increased muscle protein in individual cells, not the formation of new cells. A long distance runner during an intensive training period may not look different on the outside, but they too will have increased muscle mass, this time focused on an enlarged heart—the so-called 'athlete's heart' with increased weight, volume and chamber size. But again, this is mostly due to an increased size of individual cells, rather than new structures. The heart generally returns to its pre-exercise level once training stops.

To get real long-term differences in body structure it seems you need to be born that way or start 'treatment' at an early age. Extreme drug-induced remodelling of infants in the womb or young children has not been attempted, nor is it clear how one would go about this. Possibly very small increases in growth hormone over time might be effective in breeding a taller athlete, but it is likely that this would need to be combined with gene doping to be really effective. While parental pressure has led to some extreme examples of young children being pressured to train and perform, the level of sophistication needed (and huge risk involved) in altering a child's physical structure chemically makes this a very unlikely occurrence to envisage. In this case, at least for the child's sake, society's negative views about drugs are likely to act as a serious deterrent to most parents.

The problem of skill and team sports

Judging the limits of human performance in many sports is complicated. It is easy to determine performance increases and potential limits in running, jumping and throwing where we can measure quantitatively against an absolute parameter. We know that Usain Bolt was a faster runner than Jesse Owens. But in many sports this kind of quantitative assessment is more difficult. Of course this is part of the fun of being a spectator. Was the 1970 Brazilian World Cup winning football team (Pele, Jazinho, Carlos Alberta) the world's best ever or was it eclipsed by Manchester United's triple winning 1999 team (David Beckham, Roy Keane, Peter Schmeichel)? Baseball's 1927 New York Yankees or the 1975 Cincinnati Reds? We will all have our views on that, but it is very difficult to objectively judge sports performances in complex sports that have technical skills that are not readily quantifiable. The best we can do is use our qualitative skills. For example if performance were random, with no systematic improvement with time, it must be true that, the longer the judging period, the more likely you are to see a better team. So we should be rightly suspicious of lists that are biased towards an earlier 'golden era' of sport. Some baseball lists seem to fall into this category, claiming that 60 per cent of the best teams ever to play the sport were playing between the years of 1906 and 1934.[7]

Apparently robust statistics such as baseball home runs or hitting streaks are difficult to use as they reflect the abilities of both teams—if both pitcher and hitter increased their abilities the result might be no observed change. But of one thing we can be sure—athletes have become fitter. The revolution in training started by Roger Bannister and others has dramatically affected the aerobic fitness of athletes. UK footballers once had an alcohol drinking and poor nutritional culture that shocked the overseas coaches who came to manage in Britain. Resistance training in the gym (even when not steroid enhanced) has dramatically increased the strength of nearly all top sports people. Even chess players can build up lactic acid when they think; the famous chess champion Gary Kasparov had as rigorous a physical training regime as many sportsmen.

We can see this fitness increase if we look at how far footballers run in a game. A time-motion study over a full season was done on the UK-based Everton Football Club in the mid 1970s; it was estimated that the players ran on average just over 8,500 metres per game.[8] This increased to 10,000 metres by the 1990s.[9] For this reason, and probably no other, the Manchester United of 1999 would have beaten the Brazil of 1970, possibly by scoring a couple of goals in the last minutes of extended extra time as the opposition tired.

Most team sports require a certain minimum level of fitness. What is less clear is whether the optimum performance itself is limited by this fitness or, as seems more likely, by the technical skills required. This is a complex question and is likely to be sport dependent. Clearly baseball players and American footballers who have admitted to taking anabolic steroids or steroid-like molecules, feel that their home run prowess or strength of tackling benefits. But in the round ball version of football drug tests generally reveal recreational drugs (cannabis, cocaine) that are presumably not taken for any perceived performance benefit.

If we assume that performances in these technical sports are not limited by raw power or endurance, are there still performance-enhancing drugs that might be beneficial? How could we test this? In some technical sports there is an absolute quantifiable performance marker, just as the 100 m time is for Usain Bolt. Darts and snooker spring to mind as possibilities here. Both sports have seen significant performance increases in recent years. We can see this most easily in darts where there is no obvious interference from the opponent—every time a player has exactly the same shot to make. The maximum theoretical score per three arrows is 180. For the top twenty players in the world the average in 1998 and 1999 was 90. Ten years later this had increased significantly to 94. This is too great to be accounted for by technical changes in the game during this period which only consisted of minor modifications to prevent darts bouncing off the target.

In snooker it is less easy to find a clear statistic that is unaffected by the state of the game or the opponent's play. Perhaps the best example we can use is the frequency of the 147 break. This maximum score can only be achieved by potting thirty-six balls in order. All players strive to do this given the chance and the opponent cannot interfere once the break is started; they must remain seated passively throughout. The frequency of this rare event has increased notably over the last 10–15 years (see Figure 5), again suggesting a significant performance enhancement.

Although an increase in money has brought an increase in the player base in both sports much of the enhancement preceded this increase in the number of participants. Indeed in darts several of the same players feature ten years later—but their averages are now much enhanced. In both darts and snooker the reasons for these increases are likely to be complex and not drug related. Yet there have been some positive drugs tests e.g. the darts player Robbie Green who was banned for eight weeks by the Darts Regulation Authority for using marijuana in 2006. Although there is no suggestion that Green was attempting to get a performance benefit could a drug really have a positive affect in such a highly technical event? Cannabis, beta-blockers and even alcohol

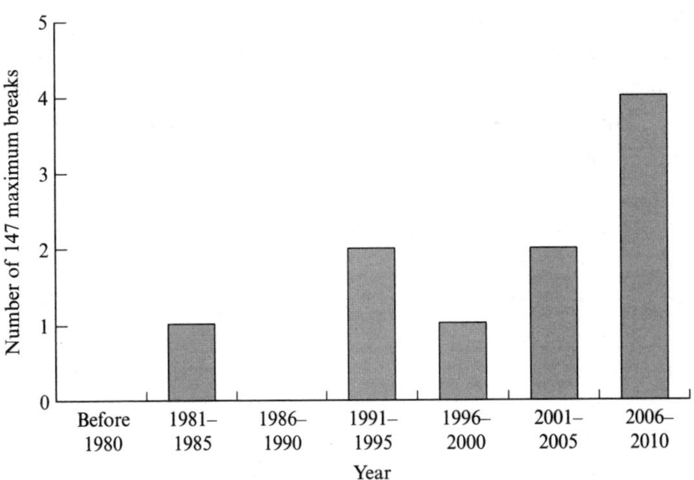

Figure 5 Distribution of 147 maximum breaks in the world snooker championships.

(the traditional, though now banned, drug of choice for darts) have all been proposed to relax sportspeople enabling them to perform better under pressure. For highly technical Olympic events such as archery and shooting these compounds are banned. But there are no anti-doping rules in the world of classical music where beta-blockers are used to lower heart rate, blood pressure and reduce stress associated with stage fright.[10] At low doses studies have confirmed this positive effect on musical performance.[11]

Drugs such as beta-blockers have the potential to enhance performance by relaxing the athlete; however, if the sport also has a significant physical component their overall effect is likely to be inhibitory. The key sport in this regard is biathlon, where athletes have to combine cross-country skiing and shooting. The art is to calm down the heart rate and focus on a highly technical event when physically exhausted. Any drug that could compose an athlete to make the killer shot would probably inhibit their aerobic skiing performance. Football doping is likely to suffer from similar problems.

One ray of 'hope' on the horizon for the dopers lies in the new classes of brain stimulants that enhance cognitive function without limiting aerobic performance. Modafinil for example is an anti-narcolepsy drug that is—allegedly—used by armies to keep their soldiers awake and alert without the problems of amphetamines. It is claimed to enhance cognitive function as well. It is only a bit fanciful to think of players remaining mentally alert, by replacing the half-time oranges (or now rather fitness drinks) with an undetectable brain stimulant. I know of no studies looking at modafinil effects on darts (or even chess) players, but this seems an obvious study to do, given its relatively mild side effects.

Biochemical limitations

Not surprisingly it is at the biochemical end of human performance where drugs and ergogenic aids really start to impact. Indeed much of the rest of the book will be devoted to this very issue. However, before we explore impacts on optimum performance we should be aware that many athletes are like the rest of us—they can make mistakes in their basic biochemical intake that have impacts on their health, and hence performance. We have noted the poor nutritional state of the average 1970s footballer; however, other more scientific sports are not immune. As a young athlete, Paula Radcliffe found herself struggling with fitness in the 1991 season. Eventually she was diagnosed with anaemia—a lack of iron limiting her ability to transport oxygen around the body. While still being able to run much faster than the average person, she was unable to compete at her peak. Following iron supplementation she was able to win the World Cross Country championships the following year and ten years later achieved multiple world records in the marathon.

Assuming that an athlete has avoided such performance roadblocks the challenge is then to optimise the body's biochemistry. Overall energy content is unimportant. What matters in biochemical terms can be expressed in terms of power. Having large fat stores of energy will not benefit you (at least biochemically, there are obvious biomechanical benefits for Sumo wrestlers). Instead the key is the ability to deliver your energy in as short a time as possible. This is what power is—energy divided by time. The atom bomb dropped on Hiroshima had an energy of 15,000,000,000,000 Calories, equivalent to every member of the UK population eating a chocolate bar. The key difference is that in an atom bomb the energy is expended in a very short time, so the power is much higher.

The ability to create a high power level—and sustain it longer than your opponent without fatigue—is the key to how biochemistry drives sports performance. In sports such as weightlifting or sprinting the requirement is very much for the highest power the body is capable of delivering. In other sports, such as long-distance running or rowing, the ability to maintain a slightly lower power, but over a much longer time period is the key. In both cases energy must be used to drive muscle contraction. This energy comes from a molecule called ATP (Adenosine-5'-triphosphate), which needs to be continually synthesised as it is devoured by the requirements of the muscle. The mechanism the body uses to deliver ATP over a very short period is different to that required for maintaining a lower power over a long period. As we shall see in the upcoming chapters this results in fundamentally different requirements for ergogenic aids and performance enhancing drugs in power and endurance sports.

Stoking The Engine: Food

'C'est la soupe qui fait le soldat'.
Napoleon Bonaparte (attr.)

Training enables an athlete to be stronger and faster; it increases the potential to deliver the power needed for top performance. But that power itself requires energy from two sources: the food we eat and the air we breathe. Analysing the biochemical means of optimising these fuel sources is the subject of the next two chapters.

Feats of long distance running are burnt into our folklore. In Western culture we honour the mythical run of Pheidippides from Marathon to Athens; he just had time to inform the Greek people of their victory over the Persians before abruptly dropping dead. This story became the inspiration for the invention by Michel Bréal and Baron Pierre de Coubertin of the only Olympic event named after a town. In Eastern culture we have the Japanese version of the 'Pony Express'—the famous running postmen who delivered mail in relays across the country. As we learnt in the introduction a long distance race between a human and a horse is a real contest. With international sport in its infancy at the time we were denied the chance for a head-to-head race between the US and Japanese pre-industrial postal systems. And rather sadly, this romantic age of postal delivery only lasted for a few decades before the arrival of the steam train on both continents.

Aerobic exercise, as exemplified by Pheidippides, is exercise in the presence of oxygen. To run a perfect 10,000 m race or marathon requires hours, days, months and even years of hard training. This training builds up your muscles, increases your heart's ability to pump blood and enhances the efficiency of metabolic fuel use in the body. But, on top of that, all athletes prepare very carefully immediately before and during their events in terms of their molecular intake. The days before a race, and the hours during it, are key times when nutritional aids and drugs can intervene to optimise human performance.

The Energy Of Life

The energy for running a marathon comes predominantly from the chemical burning of the body's fat and carbohydrate reserves by oxygen. If you take a lump of fat or sugar—a chocolate bar is probably the best combination of the two—and put it in a fire, it will burn. In this process oxygen is converted into carbon dioxide and water. This process transforms one source of energy (the chemical bonds in the chocolate) into another (more heat in the fire). But this is not of any great use. Inside the body the dramatic 'one shot' burning of this fat and carbohydrate is broken down into lots of smaller chemical reactions. By breaking up the molecules in a more structured and controlled way energy can be converted into a useable form. Confusingly the molecular process of oxygen utilisation shares a name—respiration—with the physiological process we all know as breathing. This confuses even the more scientifically trained of us; I recall having a conversation with an anaesthetist and becoming increasingly amazed at his ignorance of my beautifully reasoned arguments, before simultaneously we realised he was talking about lungs and I was talking about cells.

The key output of molecular respiration is a molecule called adenosine triphosphate (ATP). This is considered the universal energy currency of the cell. What does this phrase mean? We use ATP to drive chemical reactions in the opposite direction to where they would normally go. It is the molecular equivalent of driving a car uphill. The body couples any chemical reaction it wants to make happen to the simultaneous use of ATP. If you have a lot of ATP you can do anything. This is far more efficient than having a multitude of separate energy sources. Hence to the body having a lot of ATP is equivalent to you having a lot of money in your pocket. It can catalyse (buy) anything in principle, as long as there is enough of it to go around.

It is important to realise that there is nothing magically 'high energy' about the molecular structure of ATP. A myth has grown up, pedalled in many sports science, physiology and even (to their shame) some biochemistry textbooks that ATP is a 'high energy' molecule. In particular that it has a 'high energy' bond that drives muscle contraction.

The 'tri' in triphosphate means that three phosphate bonds are stuck on to an adenosine molecule (see Figure 6). Muscle contraction does indeed use the energy from the breaking of the terminal bond. This bond is then reformed by respiration. But the point of the third bond is not that it is intrinsically 'high energy'. In fact, as phosphate bonds go, it is in the 'Goldilocks' zone: not too strong that you can't make it from breaking down food, not too weak that, once made, it can't drive unfavourable chemical reactions. Just right for a universal energy transducer in the body in fact. Think of a chain on a bicycle. The pedal provides the ultimate source of energy and the wheel turns round and does the work. But the whole thing doesn't work without a chain to couple the two processes. The energy is not in the chain—the chain is the medium of energy transduction. Life has evolved to use ATP as the universal chain for coupling its many and varied energy-yielding and energy-demanding processes. The third phosphate bond in ATP is indeed 'magical' but

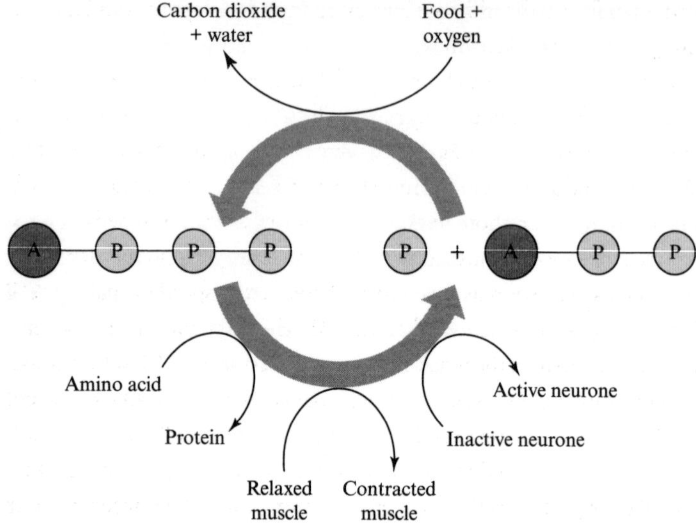

Figure 6 The cycle of life

ADP consists of an Adenosine (A) linked to two phosphate groups (P). In the presence of a third phosphate molecule, an equilibrium exists whereby some of the phosphate is bound to ADP, making ATP. Energy provided by the oxidation of food to carbon dioxide and water ensures that this reaction is favoured i.e. there is a lot more ATP than ADP. Energy is therefore temporarily stored in the increased ratio of ATP to ADP. This energy can then be used to drive a wide variety of unfavourable chemical reactions, from protein synthesis to muscle contraction to neurotransmission. The ADP that is consequently formed is re-converted to ATP using the energy from food oxidation, completing the 'energy cycle' of life. ATP is therefore a flexible energy currency, rather than a high-energy molecule per se.

not for any intrinsic energy within it, but for its adaptability to react with different energy-requiring proteins.

Now think of the bicycle being hooked up to an electricity generator. Just as electricity can be used for a variety of tasks in the house, in the body ATP formation and removal can couple a wide variety of processes such as cell growth, cell division, nerve transmission, synthesis of new molecules, removal of toxins and, most importantly for our discussion, the movement of molecules that cause a muscle to contract.

ATP therefore is the ultimate energy conduit for all your walking, running, thinking and growing.

How does this relate to human performance? Every day even the more sedentary of us use about 45 kg of ATP. But the ATP content of the body at any one time is only 0.25 kg. If ATP were not being continually re-synthesised, the total amount in the muscle would last about 1 second for someone running a marathon. In our bicycle analogy, if we stop pedalling the bicycle stops moving; just so with ATP, the energy cannot be stored but must be used as soon as it is made. Therefore the speed of making ATP limits the speed of reactions that use ATP. The pedal must be pushed fast for the chain to turn the wheel fast. You need to make ATP fast to run fast.

Interestingly there is one chemical that is used specifically to prevent ATP formation during exercise. This is dinitrophenol (DNP). This molecule allows enhanced oxygen consumption with the production of no useful energy. It does this by short-circuiting the link between exercise and ATP formation, the equivalent of cycling when the chain has come off the sprocket. So why is it used? DNP was marketed as a wonder slimming drug in the early part of the twentiethth century, with over 100,000 people being treated in the 1930s. Taking DNP converts the energy in food directly to heat, even when you are taking minimal exercise. It is one of the few drugs that undoubtedly reduces body mass, predominantly fat, with no change in diet required. Unfortunately the therapeutic window is small and the effect of an overdose is a dramatic reduction of the body's energy stores—a potentially catastrophic side-effect. DNP therefore had the dubious honour of being the first drug to be banned by the newly formed US Food and Drug Administration in 1938. Still it remains a very useful drug for any 'sport' that is concerned about visible fat, but not athletic performance. Bodybuilders are known to use DNP to get the final 'cut' look, aiming for a complete absence of surface body fat. DNP undoubtedly works, but has some of the most immediate and extreme side-effects of any drug used by body builders. The mildest of these are bad breath, excessive sweating and extreme thirst; the most severe include irreversible damage to the heart, kidneys, liver, and lungs, paralysis and, ultimately, death.

Figure 7 illustrates the different processes that the body can use to make ATP. The most efficient way to make ATP is by completely burning fat or carbohydrate to carbon dioxide and water. Depending on conditions as many as thirty molecules of ATP can be made from each molecule of sugar (glucose) or its fat equivalent. This process requires oxygen. In the absence of oxygen only two molecules of ATP are made. But these two molecules are made much more quickly as the biochemical production pathway is shorter. The amount of energy that can be made in a given time is therefore much higher in the absence of oxygen. Energy divided by time is power. The shorter time you consume your energy, the higher the power—witness the difference in getting from 0 to 60 miles per hour for a high power sports car versus the low power family saloon car.

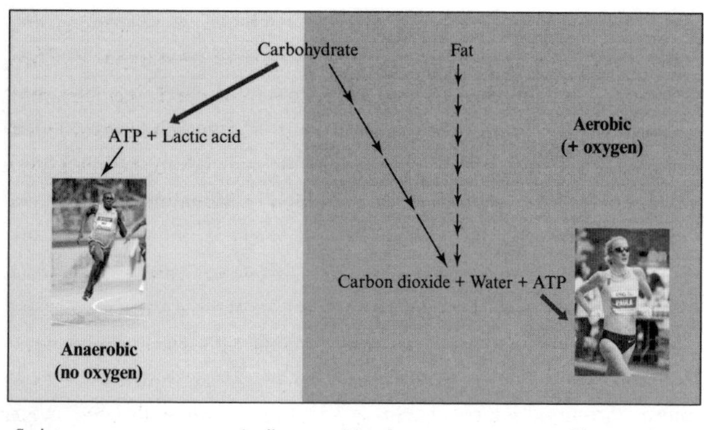

Figure 7 Different fuels for different distances

ATP is the energy source for muscle contraction. The largest fuel reserve is the process of burning carbohydrates and fat with oxygen, but this take the longest time to make ATP. Short intense exercise requires the fast, but inefficient, production of ATP in the absence of oxygen. In this process lactic acid is formed.

The pathways in the absence of oxygen (anaerobic) are therefore high power, whereas those in its presence (aerobic) are low power. These pathways are also separated spatially in cells. The high power, inefficient anaerobic pathway resides in the body of the cell (the cytoplasm); the low power efficient pathway resides in mitochondria, small sausage-shaped 'organelles' inside cells that consume all the oxygen. About three billion years ago our ancestors were large bacterial cells not able to make use of oxygen as an energy source. At the same time a smaller bacterium existed that was able to consume oxygen efficiently to make ATP. The larger bacteria engulfed the smaller one; the fusion of these two different types of bacteria has persisted to the present day. This symbiotic relationship is fundamental to all complex multicellular life; our mitochondria are the remnants of the small oxygen-consuming bacterium.

The biochemical division between the oxygen requiring low power pathway and the high power pathway is matched by an anatomical division. The muscle makes two different kinds of cells—fast twitch fibres and slow twitch fibres. Although all the cells in our body have the mitochondrial and non-mitochondrial pathways, fast twitch fibres are more adapted to the high power pathway and slow twitch to the low power. Sprinters therefore tend to have more high twitch fibres and long distance runners more slow twitch fibres.

The existence of a high power pathway is why you can run fast for 100 metres. But you can't keep this speed up for long—partly because you run out of energy and partly because you build up toxic intermediates—and you need to revert to the use of oxygen. Exercise with oxygen is more efficient; if you try squeezing a ball repeatedly with a tourniquet wrapped around your arm you will soon realise how much more quickly you tire (but don't do this at home as I can confirm it *will* hurt a lot).

In these longer distance events biochemists have developed tools to improve the supply of the fuel (fat or carbohydrate) and the gas (oxygen) that burns the fuel—the ultimate goal being to fine-tune the body's biochemistry to increase its power output. We will learn about improving the gas supply in the next chapter. What about the fuel itself?

Improving the fuel supply—fats versus carbohydrates

When it comes to using oxygen there are preferred fuels. While the overall efficiency of using carbohydrate or fat as a respiratory energy source is about equal, the biochemical pathways for fat metabolism are much slower. Therefore carbohydrates are the preferred aerobic energy source for athletes. But, even the most trim, elite athlete has far more energy in fat reserves than carbohydrates. Ultimately once you get to long distances you have to turn to these reserves. It is like having two fuel tanks in the body—one (carbohydrate) that is limited but enables high performance and the other (fat) that is unlimited but lower performing. The classic example of this problem is the naïve marathon runner who sets out too fast using only his carbohydrate energy source. They then 'hit the wall' at about twenty miles when there is a dramatic need to switch to fats as an energy source. In the words of David Costill,[1] a well-known physiologist running his first marathon 'After eighteen miles, the sensations of exhaustion were unlike anything I had ever experienced. I could not run, walk or stand, and even found sitting a bit strenuous'. Most of the training for a marathon is aimed at the (unconscious) adaptation of the body to manage the balance of fuels effectively so as to not end up like this.

So different molecular power sources are needed for different running events. Man, the 'persistent hunter' is by nature an ultra-marathoner, requiring mostly fat as a fuel. But at shorter distance aerobic events it is clear that any way you can optimise your carbohydrate energy sources, or accelerate the use of fats, will result in enhanced performance. Can we do this with nutrition or drugs?

The answer is a resounding yes. In shorter events (under the marathon) it is vital to optimise the body's use of the fast carbohydrate stores not the slow fat ones. In theory you can do this by manipulating your insulin levels. Glucose is the simplest carbohydrate in the body and is the means by which carbohydrates are transported around the body. Insulin is the hormone that controls our blood glucose levels. A lack of insulin signals the body as starving of carbohydrate; carbohydrate is therefore released from stores (generally the liver) and appears as glucose in the blood. This is why diabetics who forget to take the additional insulin they need end up in a high glucose coma; the body is tricked into thinking it is deprived of carbohydrate and floods the blood with glucose to compensate.

The opposite state exists if insulin levels are high. This indicates a body well stocked with carbohydrate. Therefore the carbohydrate stores are kept shut. In extremis blood glucose levels drop dramatically. This is why a diabetic who takes too much insulin will end up in a low glucose coma (it is also why injecting insulin is toxic).

Can athletes make use of this fact to optimise their energy efficiency? Eating a carbohydrate-rich meal an hour or so before a race can raise the insulin levels slightly. This tells the body that it is carbohydrate full, signals a happy fed state and suggests that there is no need to open carbohydrate reserves. Studies in the 1970s suggested that this could be a bad thing as upon starting to exercise blood glucose could fall precipitously.[2] However, later studies conflicted with this, and in fact suggest that the one-hour pre-exercise carbohydrate meal might be beneficial.[3] How does this work? One theory is that fat metabolism is slower in the insulin-rich 'fed state' resulting in a metabolic bias in favour of the more efficient carbohydrate fuel sources. But this is a tricky balance which athletes address mostly by trial and error. The key seems to be to raise the insulin slowly. For those who can stomach it, taking the carbohydrates in the form of lentils seems to outperform using potatoes or a glucose drink.

Although on the World Anti-Doping Agency (WADA) banned list, insulin doping is difficult to detect. There is therefore a potential attraction to adjust your insulin levels more precisely by injecting the drug and monitoring your blood glucose. Combined with just the right pre-race carbohydrate meal this could theoretically optimise performance. Although the process of taking insulin is easy enough, it seems unlikely that athletes have ever tried this. The act of going for a sneaky injection an hour before an Olympic race might be difficult to conceal. Insulin abuse is therefore probably restricted to off-season training where it has been suggested to enhance recovery after strenuous exercise and hence build up muscle mass.

Once you get to longer distances such as the marathon you really do need both your fat and carbohydrate supplies. Carbohydrate is stored in animals in the form of glycogen (in plants the equivalent store is starch). Hence it is key to have as much glycogen in your muscles as possible so that in the race you have an optimal supply of high power fuel. We know all this due to the sacrifice of a group of Swedish firemen who cycled to exhaustion and allowed scientists to take blood samples and, more painfully, muscle biopsies during the test.[4] Their last pedal correlated well with the removal of the last glycogen stores from their muscles.

Increasing these glycogen stores can be done by diet—a process called carbohydrate or 'carbo' loading. The London Marathon used to host the world's largest spaghetti party the night before the race. But, of course there are ways of doing this more scientifically. The classical method—developed by Swedish scientists in 1967—is to run a long race a week before the marathon to deplete all your muscle glycogen stores.[5] Then you go on a low carbohydrate diet for a few days and only do very light exercise (a process called tapering). The muscles are now desperate for a 'glycogen fix'—or to put it more scientifically the biochemical mechanisms to optimise glycogen storage in the muscle have been upregulated. Then for the few days before the race you eat as much carbohydrate as you can. There is a supercompensation and muscle glycogen rises to levels higher than ever before. The race starts and the athlete can run longer with this enhanced reserve of high power fuel. Later adjustments to this regime modified somewhat this classical protocol[6]—athletes complained the very low carbohydrate phase gave them diarrhoea—but the general principle has remained the same from the 1960s to the present day.

Like many methods that will be discussed in this book, these nutritional effects are most dramatic on the untrained person. An elite marathon runner will already have very high glycogen stores and the supercompensation effect is smaller. But in the untrained person the effects can be dramatic—more than doubling glycogen stores and increasing competitiveness.

What about during the race itself? Can drugs or nutrition help? In terms of fluid intake there is no reason to do anything much in intense exercise lasting less than sixty minutes, except perhaps drinking water on hot days to prevent dehydration. But in longer events taking nutrients on board can make a big difference. Sports drinks, such as Lucozade™ or Gatorade™, really do what the adverts say and keep you going longer[7]—though it has to be said, sponsorship issues aside, most elite athletes generally make up their own specialised drinks. What are the key components in these drinks and how do they work? The technical term is carbohydrate-electrolyte (CES) solutions. The electrolytes replace the salt lost in sweating, whilst the carbohydrate improves the energy supply. After sixty minutes running, blood glucose levels start to fall and drinking CES solutions prevent this. A fall in blood glucose is bad for two reasons; first it is a key fuel for the brain and the last thing you want is to send a signal to the brain, consciously or unconsciously, that what you are doing is bad. But a fall in blood glucose also signals carbohydrate depletion, resulting in a decrease in insulin levels that triggers the undesirable release of the low power fat stores. The benefits may not only relate to marathon runners. Football players have also been shown to decrease their muscle glycogen during a match. There is a good scientific rationale to the move away from the pre-match high-protein steak diet to the pasta of today's top players; and although the half time oranges of yesteryear will do some good, a well-balanced CES sports drink is probably a better bet.

There is also a good scientific rationale for 'carbo-loading'. However, fat loading is more controversial. In this case the strategy is more long-term. Moving to a high fat diet is alleged to lead to adaptations in the body to improve the rate of fat metabolism. The idea is that these adaptations carry over into exercise, resulting in improvements in the longer distance races where fat must be used. Indeed in rats this is exactly what happens. Rats fed a high-fat/low-carbohydrate diet were able to run for longer than those on a normal diet.[8]

Many of the changes in fat metabolism seen in rats after the high fat diet are the same as those seen following endurance training in humans. However, before you run away with the idea that ice cream could be the major component of your new marathon training programme, this work has not been readily transferable into humans. Training on a high fat diet does seem to result in increased fat oxidation ability, but this is balanced by a drop in glycogen stores; the end result—no effect on performance.[9] Athletes are currently trying to balance training on a high fat diet with last minute carbohydrate loading to get the best of both worlds. It may be there is a secret in the exact timing of the dietary switch.[10]

There has been a recent intriguing take on the carbohydrate drink. Researchers at the University of Birmingham in the UK have shown that rinsing alone with a carbohydrate mouthwash can improve performance in long distance laboratory time trials. This effect clearly cannot be due to energy use; it must have a neural component. Indeed studies suggest that there is a specific part of the brain that is activated when sugar solutions are in the mouth.[11]

Carbohydrate loading and sports drinks are clearly nutritional tricks. The science outlining the problem and the testing of possible solutions may be high tech; however, the solutions themselves—high carbohydrate diets and sugary, salty drinks—would have been available to man throughout evolution. There is no particular evidence that these strategies have been employed on the evolutionary timescale, but perhaps that is not surprising. The marathon is the race distance at which these nutritional tricks work best. But hunting strategies do not require continuous running for two hours. As noted in the last chapter the persistent hunter walks and runs for many hours or days. The insulin system is perfectly capable of optimising the balance between fat and carbohydrate for this kind of performance. So the dietary needs are a bit less specialised. Whilst there is a debate about whether it is beneficial to add proteins to carbohydrate sports drinks for 'short' events like marathons, such mixed fuel drinks are commonplace for ultramarathoners. Their fuel consists of carbohydrate, salt and protein; in short the sort of food a hunter might anyway take with him on the hunt whether for a short trip or an extended multi-day hunt.

Pre Race Tonics

It is difficult to think of an evolutionary situation where being able to run as fast as you can for 26 miles 385 yards, or alternatively being able to kick a round ball made of animal skin for ninety minutes, would be of much benefit. So it is perhaps not surprising that using the tricks above we can optimise nutrition to outperform our evolutionary constraints. But what about more specialised molecules unknown to previous generations? Are there super drugs that can do even better than nutritional aids? For the budding superhuman the answer to these questions is a little depressing.

THE THREE C'S: CREATINE, CARNITINE AND CAFFEINE

Creatine is a small molecule that we make in our body, but is also a component of our diet, especially in meat. Being a normal part of our diet the anti-doping authorities have not banned it, so we know quite a lot about its use in sports. It was fashionable in the 1990s and for a while was de rigueur in many top football clubs in Europe. In England it even led to an argument between the new 'scientific' approach of the Arsenal manager, Frenchman Arsene Wenger and the more traditional methods of the Scottish Manchester United manager, Sir Alex Ferguson. Arsenal eventually stopped their creatine experiment after players complained of stomach cramps.

Creatine pills can include over five times the normal meat-eaters' daily intake; cramping is indeed a relatively common reported side effect, though this has been difficult to reproduce in laboratory testing conditions. How does creatine work? Essentially creatine loading increases the resting muscle content of not only creatine, but a related molecule called phosphocreatine. During intense exercise phosphocreatine is converted to creatine, generating a short additional burst of ATP production. In longer-term aerobic exercise this has no effect, but it seems to be of benefit in short-term repeated sprints.[12] So the attraction to football managers can readily be seen. Creatine is also used during training to build up muscle mass—the mechanism of action here is more obscure and does not seem to relate directly to energy metabolism.

Carnitine, like creatine, is a molecule that we make in our body, or ingest in our diet. It plays the role of transporting fat molecules into the mitochondria where they are subsequently broken down to be consumed by oxygen and produce ATP. Increasing carnitine levels could theoretically improve the rate of fat transport to mitochondria. This would increase the power of the fat oxidation pathway that limits the rate of running in a marathon or ultramarathon. The problem is that taking carnitine pills—or even infusing carnitine via a drip—does not seem to increase carnitine in the human muscle.

The transport of carnitine from the blood into the muscle is very slow. So the lack of reported effects on performance is hardly surprising.[13] However, increasing carnitine uptake is one of the ways insulin enhances fat oxidation. So it should be possible to increase carnitine levels by altering blood insulin. Modifying insulin levels is most effectively done by taking the drug itself. However, athletes are currently trying to achieve a similar effect without risking a drugs ban by taking high doses of carbohydrate at the same time as they take the carnitine. The idea is that the carnitine pill increases levels of carnitine in the blood whilst the carbohydrate-induced increase in insulin levels enhances its transport into the muscle.

Caffeine is a different biochemical beast altogether. Like insulin and creatine it will appear in a number of chapters in this book. As a student I used to fear having to learn about these compounds, so many and varied are the biochemical pathways in the body with which they can interact. Yet herein lies the attraction to the doper. One pill can have multiple effects. Initially it was thought that caffeine directly increased the rate of fat oxidation—the same idea proposed for carnitine. However, it now seems likely that caffeine has little effect on energy metabolism. Taking it before a race will give you an advantage, but not in fuel metabolism.[14] Instead it now seems more likely that it works directly as a stimulant of the central nervous system. It is most effective when taken an hour before exercise bouts of between five and thirty minutes; two strong cups of coffee, a bottle of a caffeine-based energy drink or a caffeine pill, are enough to make the difference.

BICARBONATE AND CITRATE: THE TAMING OF ACID

Bicarbonate is a compound that most people will have used in their daily life either as baking soda in making cakes or as an antacid to overcome indigestion. This dual function is because bicarbonate gives off carbon dioxide (your cake rises) and absorbs acid (your stomach feels better). Why is it used by many athletes before a race? The perception is that it can solve problems with muscle acid production during exercise. It is therefore the antacid, rather than the baking soda effect that drives its purported performance enhancing—or ergogenic—use.

Lactic acid builds up over short distance running. It is the acid part—the hydrogen ion or proton—not the 'lactic' that is the problem here. Acid has the potential to damage the proteins in the cell that drive muscle contraction. This balance between acid and alkali is given the term pH. The pH scale varies from acid (pH 0) to neutral (pH 7) to alkaline (pH 14). Equal concentrations of acid and alkali are neutral (pH 7). This is the pH of completely pure water. A lower pH is more acid and a higher pH more alkaline. As any child knows who has brought home colour pH indicator paper from school, the red wine vinegar in your home is acid with a pH of about 2.5; likewise a glass of orange juice comes in at 4.5. On the alkaline side, milk of magnesia is 10 and household bleach 12.5.

The pH in most body cells is close to neutral. Exercise can cause the pH to fall as acid builds up; in the most extreme cases the drop in pH in the muscle cell is from 7.1 to 6.5. At first sight the change might appear small. However, pH is defined mathematically as the negative logarithm of the concentration of protons. This means the proton concentration rises exponentially as pH drops; the apparently small change in exercise in fact equates to a three-fold increase in the concentration of potentially reactive protons.

What effect does this drop in pH have on muscle function? There are many potential targets as a wide range of chemical reactions in the body use or consume protons. However, the role of pH is even more ubiquitous. The body uses catalysts to control chemical reactions. These catalysts (termed enzymes) are protein molecules. Proteins are made up of tens to hundreds of building blocks called amino acids; it is the sequence of these amino acids that is coded for by our DNA and characterises the specific function of the protein. But amino acids, the building blocks of proteins, are able to react with acid or alkali. So a chemical reaction does not itself have to involve a proton to be affected by acidity; the enzyme catalyst can be altered by the pH change. In most cases the pH at which this happens is in the far acid (< 3) or basic (> 9) end of the pH scale. But a significant number of amino acids react to changes around the neutral pH range within which our body operates. The upshot of this is that any enzyme reaction will be affected by pH to some extent, even if a proton is not directly involved in the chemistry.

Nevertheless muscles seem somewhat resistant to changes in pH. Although acid pH does inhibit key individual enzymes in the pathway of carbohydrate breakdown, the overall rate seems relatively unaffected. Indeed, in the test tube, isolated muscle cells contract happily in the presence of higher lactate and lower pH than occurs in the body during the most intense exercise.[15] Yet it is clear that the lowered pH and increased lactate during exercise correlate with fatigue. Either this correlation is not causal or there are other things going on in the whole body that we do not know about. Increases in potassium and phosphate or disruptions in calcium movement in cells have all been suggested as alternative inducers of fatigue.

Be that as it may, the historical view that acid inhibits muscle function has led to the attempt to use molecules in the body that stop this acid forming. One of these is bicarbonate, which removes acid, making carbon dioxide and water in the process; the carbon dioxide gas produced is then breathed out. Eating capsules of bicarbonate two to three hours before strenuous exercise seems to improve performance in events lasting between a minute and five minutes i.e. exactly those where there is a combination of aerobic and anaerobic exercise and you might expect lactic acid build up. The downside is perhaps predictable from anyone who has drunk too much of a fizzy drink. But in the case of the carbon dioxide doses involved the effects go beyond mere belching and flatulence; bicarbonate loading can lead to vomiting and diarrhoea, symptoms likely to limit its positive effect on performance.

For this reason, some athletes prefer the use of different molecules that can mop up the acid without the deleterious side effects. Citrate is a possible solution. However, it too is not immune to negative effects on the gastrointestinal system. In a 2008 study fifteen of seventeen athletes reported mild nausea, stomach cramps, cold shivers or diarrhoea after taking sodium citrate pills.[16] Interestingly most people will have already used citrate to balance a change in pH. It is the active ingredient in 'pH balanced' shampoo. The citrate is added at an acid pH (in the form of citric acid) and resists any change in that pH (about 5.5) Therefore pH balanced really means acidic, but of course this doesn't sound so good in the adverts. Acid pH is supposed to be slightly better for the hair as it mimics the hair's own pH (about 5). However, as the shampoo is washed out quickly and citrate is only slowly absorbed, it seems unlikely that pH balanced shampoos—or their fierce rivals the 'ultra neutral' shampoos—really make any difference.

So do bicarbonate and citrate really work to enhance sporting performance? Proponents claim that under the right conditions they can have a positive effect for races between 400 m and 1500 m, where lactic acid production is expected to be high. Under laboratory conditions studies are promising. But the effect on athletes in real races is more equivocal. The mechanism of action is also not entirely clear. Bicarbonate and citrate can reduce pH changes in the blood. But when they are taken before a race, they do not appear to have an effect on pH changes in the muscle itself.[17]

Spiking Your Energy Drinks

DRINKING THE FAT

Is it worth spiking the energy drink you take during a race with more than just salt and sugar? A range of exotic combinations have been tried. Some aim to optimise energy delivery and others to overcome fatigue. As the rate of fat oxidation is a clear limiting factor in any race of marathon or greater distance, many attempts have been made to enhance the rate of energy production from this source. Medium chain triglycerides (MCT) are smaller versions of the fat molecules the body normally uses for its energy reserves. The idea of taking them in a drink is that you can short circuit the circuitous path the fats take from storage cells to the mitochondria. Anything that increases the rate of fat oxidation has the potential to decrease the time taken to run a marathon or ultramarathon. The theory is great and MCTs initially seemed to work in the laboratory. However, more recent studies have shown no effect.[18] As with bicarbonate and citrate, so with MCTs the side effects are not fun; ingesting 125 g ($^1/_4$ of a pound) of fat, however small or large the molecules are, can result in nausea, vomiting, stomach cramps, bloating and diarrhoea.

Some more exotic methods have been tested to enhance the rate of fat oxidation. Patients with diseases that impair their ability to consume oxygen from blood sugar are treated with intravenous infusions of fat in the form of a compound called triacylglycerol, designed to bypass the damaged glucose pathway. This has led athletes to wonder whether this enhanced fat supply could act *in addition* to a normal sugar pathway. Although triacylglycerol is slowly absorbed from the blood this can be increased by co-administration with heparin.

Heparin is a molecule naturally produced by white blood cells as an anti-coagulant. But it also has the property of activating an enzyme that can break down triacylglycerol. Triacylglycerol breakdown produces molecules called free fatty acids that are direct energy sources for the mitochondria. Combining infusions of triacylglycerol and heparin into the bloodstream during exercise can indeed alter fat metabolism. But it is hard to see how they could be mimicked in a race environment. Intravenous infusions of fluids are not permitted and of course you could not run effectively with a drip in your arm. One-off intravenous *injections* though are allowed, and heparin and triacylglycerol are not on the doping list. So as far as I can tell an athlete would not be formally disqualified for injecting themselves at a refreshment point in the middle of a race. Even if feasible, and not banned, it is easy to see the effect such an image would have on any observers. The conceptual link to recreational drug use would be too much for this to be allowed, even if the fluid was simply a fat supplement.

PROTEIN 'SHAKES' AND BRAIN FATIGUE

After carbohydrate and fat the third energy source in our diet is protein. Protein shakes are the staple drink of bodybuilders, helping to supply the building blocks for muscle growth. But is there any point taking a shake in a race? Some studies suggest that adding proteins to a sugar drink is beneficial; critics note that in these studies the sugar content of the drinks was anyway suboptimal so the studies were not a fair comparison. But proteins may not exert their influence immediately. The general consensus emerging seems to be that, while it is unclear whether adding proteins during a race is beneficial *at the time*, it can reduce muscle soreness after the race and hence enhance post-race recovery.

One specific type of protein, or rather protein breakdown product has been suggested to act in a much more specific way. Proteins are built up from amino acids. Indeed when you drink a protein shake the protein molecules are digested into their amino acid constituents before being absorbed into the body. Once in the body they are either used to build up new proteins or are broken down further to release their energy. In terms of raw energy content one amino acid is very much the same as another. But some amino acids have direct and indirect effects on brain signalling.

The body's metabolism is a web of interacting, interconverting molecules. Amino acids interact with this metabolic web at different places; they are not just simple energy sources and building blocks for proteins. Instead they can make many other molecules. In particular one amino acid, tryptophan is a precursor for serotonin. Serotonin is a neurotransmitter—a brain signalling molecule. It has the ability to regulate mood. Increasing the brain content of serotonin is a prime target of antidepressant drugs such as amitriptyline, dothiepin, fluoxetine and paroxetine. The side effect of many of these drugs is drowsiness and fatigue. Increasing your serotonin levels induces fatigue.

So how can drinking a specific amino acid prevent fatigue during exhaustive exercise? The argument was first penned in 2000 by Eric Newsholme, a marathon-running Oxford don. It is a clever idea (at least to us biochemists), and goes like this.[19] The amino acid tryptophan is the precursor to serotonin formation in the brain. Tryptophan enters the brain from the blood via a specific transport mechanism. A protein binds to tryptophan and helps it enter the brain. However, other amino acids have a similar molecular shape to tryptophan. They can therefore piggyback on the same transportation method. These are the branched chain amino acids (BCAA), leucine, isoleucine and valine. If we can increase the amount of BCAA in the blood we can therefore slow down tryptophan transport into the brain. Less tryptophan means less serotonin. Serotonin makes you tired. Ergo adding BCAA in your sports drink will make you less tired.

It is one of those theories that is too good not to be true; as a biochemist you are almost rooting for Newsholme from the sidelines. Indeed, like many things in science, the preliminary studies were promising, both on race performance and post-race mental agility. But as T. H. Huxley once quipped, the great tragedy of science is 'the slaying of a beautiful hypothesis by an ugly fact'. Laboratory studies in humans have shown that taking BCAA in sports drinks can indeed raise the blood ratio of BCAA to tryptophan; but the evidence for a performance effect is unfortunately much more equivocal.[18]

What about more extreme ways of altering the brain's signals? The serotonin fatigue argument can surely be addressed more specifically using pharmacological, rather than dietary tools. Specific drugs can be used to decrease or increase serotonin levels in the brain. Rats run for a longer time on a treadmill when their serotonin is decreased pharmacologically and get exhausted quicker when it is increased. Antidepressants such as paroxetine are designed to raise serotonin levels; anti-migraine drugs, such as pizotifen aim for the opposite effect. So far in human studies the former has been shown to reduce endurance running performance, but somewhat surprisingly the latter has had no beneficial effect.[20] This is perhaps why, whilst many brain stimulants are banned, smart pharmacological agents to modify brain fatigue have never featured on a list of prohibited substances. Personally I feel that this is an omission; if I were a doper this is an area I would definitely want to get involved in.

POPEYE, BEETROOT JUICE, AND THE EFFICIENT RUNNER

Occasionally you read a bit of research that genuinely surprises you. Frequently it doesn't stand up to closer scrutiny, but when it does it is one of the joys of being a scientist. One such case is the effect beetroot juice has recently been shown to have on long distance sports performance. The story starts with Alfred Nobel. The fortune that resources his eponymous prizes was made by selling explosives, based on nitroglycerin. However, nitroglycerin can also be used as a drug to help treat blood flow restrictions in angina. Indeed Nobel himself was treated this way. Yet the mechanism of its therapeutic action remained unknown until the end of the twentieth century, when it was discovered that nitroglycerin is converted inside the body to a gas called nitric oxide. Nitric oxide, once thought of as merely a toxic environmental pollutant, is now known to play a major role in controlling blood flow, blood volume and blood pressure in the body. In 1998 the prize that was funded by the profits of nitroglycerin as an explosive, was awarded to the scientists that discovered the mechanism of its use as a medicine.

Nitric oxide plays a normal physiological role in the body. Nitroglycerin just gives the body's synthetic pathways a boost. At first it was assumed that all of the nitric oxide in the body was produced by a specific enzyme called nitric oxide synthase. However, in 1995 a doctor from St Bartholomew's Hospital in London, Nigel Benjamin, showed that there was an additional pathway at work.[21] We eat nitrate in our diet; especially high amounts are found in vegetables such as beetroot, lettuce and spinach. The nitrate we ingest becomes concentrated in our saliva. Bacteria in our mouth have enzymes that can convert this nitrate into the related molecule called nitrite. Inside the body there are a range of chemical reactions that can turn this nitrite into nitric oxide. The acid conditions of the stomach are ideal for this and Benjamin showed that when we eat nitrate we make nitric oxide in the stomach; indeed you can actually measure the gas on your breath after you eat.

Why does this happen? Nitric oxide is toxic to the bacteria *E. coli* and *Salmonella* that live in the stomach. It has been suggested that the body has evolved systems to make use of the 'good' bacteria in the mouth to make a gas that kills the 'bad' bacteria in the stomach. As lettuce is full of nitrate, it has been more fancifully suggested that this is the reason why the salad course traditionally preceded the—potentially bacterially infested—meat course. Be that as it may, this alternative route of nitric oxide formation has become of great interest in a range of possible therapies. Nitrate pills are used as a drug in many countries and nitrate injections and inhalations are being used to control blood flow and blood pressure.

How does this relate to sport? Over ten years after Benjamin's original work, a discovery was made in Nobel's homeland, actually in the same Karolinska Institute in Stockholm that chooses the winners of his Prize. Björn Ekblom's research group found that nitrate pills could make runners use their oxygen more efficiently.[22] For the same level of exercise intensity, less oxygen was used. There are no other ergogenic aids or exercise training regimes, however extreme, that can even approach this effect of nitrate. Andy Jones—one of the sports scientists who advises Paula Radcliffe—wanted to follow this research up at the University of Exeter in the UK. But he couldn't easily use nitrate in his volunteer subjects as it is licensed as a pharmaceutical drug. Instead he did the next best thing and gave them a beetroot juice drink that contained equivalent amounts. Beetroot juice was shown to have the same positive effects as nitrate pills and the nitrate present in the juice was shown to be the active component.[23] What's more not only did beetroot juice enable oxygen to be used more efficiently, it improved performance in time trials.

Beetroot juice has also been shown to lower blood pressure in patients suffering from clinical hypertension. You don't even need to drink juice. The levels that improve sports performance or lower blood pressure are possible to reach by eating significant amounts of lettuce, spinach or beetroot in the diet—as long as you avoid those from organic farms that don't use nitrate fertiliser. There might have been something in the cartoon character Popeye's penchant for tins of spinach—at least to make him run away faster, if not to increase his strength. However, he would have had to change his eating habits. Popeye swallowed the contents of a tin of spinach in one mouthful. However, studies have shown that not chewing enough or spitting out the resultant saliva all prevent the bacteria from doing their job to convert nitrate to nitric oxide. There was indeed something in our parents' entreaties to eat vegetables, chew our food and not to spit. What we weren't told, but probably should have been, is not to use antibacterial mouthwash. This kills the good nitrate metabolising bacteria in the mouth and hence prevents nitric oxide production.

I have been working on nitric oxide and oxygen metabolism for almost twenty years and this is one of the most surprising results I have seen. Not so much in the overall effect, as the gas is known to interfere with oxygen metabolism in cell and animal studies, but that it can be so easily manipulated by a simple dietary change one hour before performance. The detailed molecular mechanism is still elusive; it is possible that nitrate may be having effects other than via manipulating nitric oxide. But the performance effect is robust; athletes have taken note.

What's In A Name: Performance-Enhancing Drug Or Nutritional Ergogenic Aid?

While researching this chapter, I was surprised how little has been done to enhance the efficiency of fuel production. With the possible exception of nitrate, successes have been largely nutritional, rather than pharmaceutical. In one sense this might appear obvious. We are what we eat and so there is clearly the potential to alter our fuel use by altering our nutrition. But, apart from the dramatic negative example of dinitrophenol—converting useful energy into heat—we seem to lack pharmaceutical tricks to optimise the body's biochemical use of energy. It is possible that we just don't know enough about the necessary subtle alterations in our metabolism. This is not for want of trying. In a number of diseases, such as cancer, AIDS and congestive heart failure there is an associated pathology due to weight loss (cachexia) and muscle wastage. There is big money in solving these problems. The muscle wastage can be dealt with in part by the use of anabolic steroids. However, pharmaceutical companies seem unable to affect the weight loss. It may just be a biochemical quirk of nature that it is easier to design molecules that can enhance muscle mass rather than those that improve fuel metabolism. If the pharmaceutical companies do start finding solutions, I would predict a rush of performance enhancing drugs in sports that require optimal aerobic metabolism. Perhaps nitrate will be the vanguard of this revolution?

Whilst no one criticizes the use of different energy fuels in food and sports drinks, the use of some nutritional supplements is not without controversy. So what is the problem? The case for the defence would appear to be that these molecules are 'natural' parts of the body's diet. But the levels of creatine and carnitine sold in pills to athletes to improve performance are way above anything that can be achieved naturally in the diet. Similar arguments can be made for citrate. What makes one natural molecule targeted at fat oxidation (carnitine) a permitted supplement, but another targeted at muscle synthesis (testosterone) a doping agent? It actually requires quite a sophisticated biochemical argument to separate these compounds. Although not explicitly stated the idea appears to be that nutritional supplements present at high concentrations that participate in bulk metabolic reactions are fine; hormones and other signalling molecules present at lower concentrations that control the rate of these reactions are banned. The exception to this rule is caffeine. It fits completely in the low concentration signalling category, is not even a natural hormone, but remains fully supported by sporting bodies and has been removed from all banned lists. As a legal, recreational drug in society, sport has given up trying to regulate its use, leading to this anomaly.

At present sporting bodies themselves have no qualms at all in promoting these supplements. The world famous Australian Institute of Sport, for example, has been operating a supplement programme for athletes since 2000. Supplements are ranked on a risk: benefit analysis. Caffeine, creatine and bicarbonate are all in group A, the top category fully supported for use.

However, there is one completely natural performance-enhancing supplement that has to be given in bulk and yet is banned by all sporting bodies. That is our blood, which does the job of carrying oxygen around the body; in the next chapter we find out what happens when the food we eat meets the oxygen we breathe.

Stoking The Engine: Oxygen*

The word aerobics came about when the gym instructors got together and said, 'If we're going to charge $30 an hour, we can't call it jumping up and down.'
after Rita Rudner

Increasing oxygen consumption is good for us all. Yet what to you and I is a desire to lose a few pounds is to the elite athlete at the very centre of their sporting performance. This chapter explores the extraordinary lengths elite athletes will go to increase their aerobic metabolism, with the ultimate aim of improving sports performance.

* Parts of this chapter were adapted from the article: Cooper, C. E. 2008 The biochemistry of drugs and doping methods used to enhance aerobic sport performance. *Essays in Biochemistry* 44, 63-83 © Chris E. Cooper.

Oxygen is carried in the blood. The heart pumps that blood around the body. When it reaches your legs or arms it is delivered to muscle cells via a network of small blood vessels (capillaries). The blood gets as close as it can to these muscle cells; then it offloads its oxygen. The oxygen travels to the mitochondria where it finally comes to the end of its journey, ending up transformed to water molecules. Just like oxygen can fan a fire to enable energy for destruction, each mitochondrion 'burns' sugar in the oxygen to enable energy for construction; in the case of an exercising muscle this is the energy that drives the contraction of muscle fibres necessary to lift, jump and run. One way of improving your exercise performance is to train hard to increase the power of your heart to deliver more blood around the body and train your muscles to deliver that blood to where it is needed (see Figure 8). This chapter will not focus on the prerequisite training required to increase cardiac output, muscle blood flow and mitochondrial content. Instead it will explore the additional physiological and pharmacological tricks aimed at supplementing this hard work.

Is blood oxygen content limiting for aerobic exercise?

The word aerobic comes from two Greek words meaning air (aero) and life (bios). The revolutionary work of Priestley and Lavoisier in the eighteenth century demonstrated that it is the oxygen gas in the air that is the necessary component for supporting life. So surely it is obvious that the maximal rate of aerobic exercise must therefore be limited by oxygen? Perhaps surprisingly this leads us into a number of controversial areas of current physiology. Although the term 'aerobic exercise' is one where the scientific and public definitions are broadly similar, we still need to define terms precisely to answer specific questions. I define aerobic exercise as exercise that is generally performed at levels when blood lactic acid levels are constant; anaerobic exercise is when lactic acid increases. The 'lactate threshold', the amount of oxygen consumed at the point where blood lactic acid starts to rise, is used as a marker of

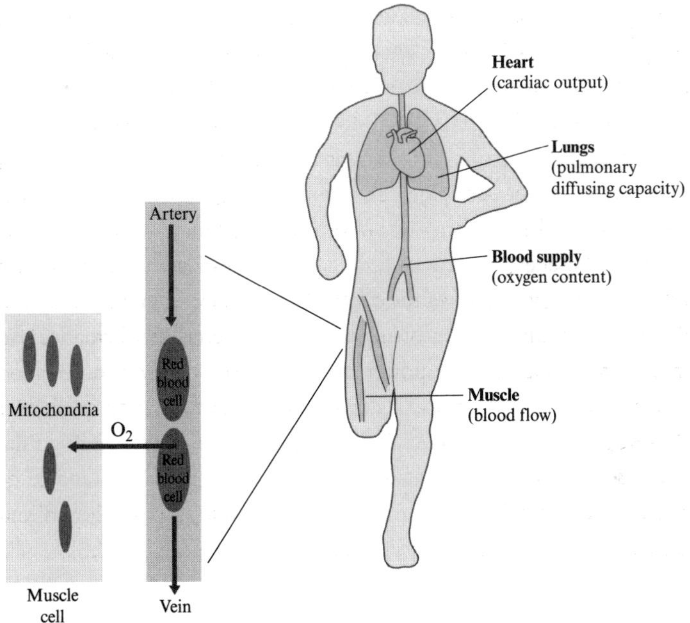

Figure 8 Factors that can control oxygen consumption during exercise

Oxygen enters the body via the lungs, is transported stored in red blood cells, pumped around the body and then diffuses from the small capillaries to the mitochondria within the muscle cell, where it is ultimately consumed. Any of these processes (indicated in the figure) could limit the whole body consumption of oxygen and hence creation of useful chemical energy.

aerobic fitness. It is improving this threshold that is credited with Lance Armstrong's Tour de France performances.

So what is lactic acid and what makes its concentration increase? Lactic acid is one of the intermediates formed when sugars are used to create useable energy to drive muscle contraction. Forming this acid creates useable energy by a process called glycolysis. Normally the rest of the energy stored in the sugar is released when the lactic acid is converted to water in the mitochondria—this is called oxidative phosphorylation. However, as already mentioned this process requires oxygen. If oxygen is in short supply then the lactic acid intermediate can build up. To get the same amount of energy the body then increases the rate of glycolysis to make up for the decrease in oxidative phosphorylation. This is termed the 'Pasteur effect', named after Louis Pasteur the famous French microbiologist who championed the germ theory of disease. Pasteur noted that the rate of fermentation in yeast increased in the absence of oxygen. In the case of yeast the fermentation produces ethanol as a by-product. Fortunately for athletes our fermentation product is lactic acid, so at least we don't have to worry about getting drunk as well as getting tired when we run out of oxygen. Lactic acid can be made so fast that it can power very high energy muscle contraction.

If exercise in the presence of oxygen is aerobic exercise then 'anaerobic exercise' should technically be exercise in its absence. However, although the dictionary definition of anaerobic is 'life without oxygen', when we talk about anaerobic exercise in most cases oxygen is still being delivered. The flow of oxygen continues maximally and oxygen is still extracted by the mitochondria. But even removing all this oxygen (making the muscle anaerobic) does not generate enough power. The mitochondrial oxidative phosphorylation system, being 'low power/high efficiency', needs to be aided by the 'high power/low efficiency' system of glycolysis. Hence the Pasteur effect kicks in and lactic acid levels rise. This is fine in the short term, such as a 100 m sprint or weight lifting, but is unsustainable in the long run; the inefficient energy production via anaerobic glycolysis results in a rapid depletion of the cell's sugar supplies and a build up of acid in the muscle and blood.

This is the same issue as when you drive a car. If you drive at 70 miles per hour (mph) to your destination you get there faster than at 50 mph. So for the same energy cost (travelling the same distance) the time is decreased. Power is energy divided by time; the 70 mph car is therefore more powerful. However, it is less efficient—you have used more fuel to get the same distance. So just like glycolysis this is unsustainable in the long run. The low power, more efficient, 50 mph car will win a race where it is the distance travelled on one fuel tank that matters. This is the modern sporting version of Aesop's hare and tortoise.

So the aim of the 'aerobic' elite athlete is to maximise mitochondrial oxygen consumption to make optimal use of the high efficiency system. The maximum rate of whole body oxygen consumption obtainable during an incremental exercise test, termed $\dot{V}O_2$ max, is a marker of aerobic fitness. Take two groups of people, one fit and one unfit, and the latter will have a lower max. It is the ability to run fast without requiring glycolysis to increase your lactic acid production that is at the heart of the success of athletes like Paula Radcliffe.

So what limits $\dot{V}O_2$ max and alters the lactate threshold? This is a surprisingly controversial question. Some groups claim that the maximal theoretical rates of oxygen delivery and consumption are not directly limiting factors per se; instead the brain acts as a 'central governor' to turn off exercise to prevent brain damage. The argument goes like this. The brain is an organ that completely requires oxygen at all times. The body is never going to let the muscles steal oxygen from the blood supply that could fuel the brain. Ergo the brain signals the muscle to stop working before this can happen even if there is still a potential reserve in oxygen available. There must be a central (brain) component to elite aerobic sports performance; otherwise there could be no psychological effects on performance, a clear *non sequitur*. However, the central governor theory goes further and claims the brain is the major factor controlling aerobic exercise performance. There are few adherents to this theory when put in these strong terms.

Yet even if the governor were only a minor component of performance, overcoming it would be a dream for every athlete. One ongoing problem is that the mechanism of the governor (physiological, let alone biochemical) is unclear. Perhaps it is best to think of the 'central governor' hypothesis as a physio–psychological phenomenon. The theory underpinning amino acid sport drinks that we explored in the last chapter would fit into this category. If brain serotonin levels rise during exercise and induce fatigue—a state that can potentially be prevented by administering drinks containing branched chain amino acids—this would be a mechanism for a biochemical event to induce a psychological feeling of fatigue.

There is clearly a rich field of neuropsychology and neurochemistry of exercise-induced fatigue waiting to be explored. However, the biochemistry of oxygen delivery and metabolism has so far proved far more fruitful to athletes and their coaches looking to optimise their aerobic exercise capacity. Getting oxygen to where it is needed—mitochondria inside muscle cells—can be broken down into a large number of stages, a simplified illustration of which was given in Figure 8. Air is normally 21 per cent oxygen (the useful stuff) and 79 per cent nitrogen (the rest). Can increasing the oxygen content of the air by breathing 100 per cent oxygen raise $\dot{V}O_2$ max? This is of course not easy to do in a sporting context; for one thing it is hard to hide the gas cylinders and they are too heavy to carry around. However, understanding why it does not have a dramatic effect on aerobic sports performance is useful when it comes to explaining what does help later on.

We get our oxygen in the blood. The blood gets its oxygen from the lungs. Blood is made up of red blood cells, a few white blood cells, platelets and plasma. The blood plasma transports many useful molecules, such as hormones, around the body but at its core it is a sweet and salty solution of water. As anyone who has tried to hold their breath in the ocean or in a cup of sweet tea can attest to, although these watery solutions have many wondrous properties for life, oxygen transport is not one of them. This is because oxygen is not very soluble in water. The more soluble a molecule is, the greater the number of molecules that can be dissolved in a liquid. So if you add two spoons of sugar to your tea it is completely soluble, but if you try to add 200 you will find that there is a lump of undissolved sugar at the bottom of your cup; the tea has become saturated with sugar.

Sugar is a solid. However gases, such as oxygen, also dissolve in solutions. But oxygen does not dissolve well in the watery plasma. Fortunately red blood cells contain haemoglobin—the protein that gives blood its red colour. Haemoglobin allows more oxygen to dissolve in the blood by directly combining with the oxygen molecules (in the process of binding oxygen the haemoglobin changes colour from claret to bright red). For every haemoglobin molecule present, up to four extra oxygen molecules can be dissolved in the blood. This ensures that the vast majority (98 per cent) of oxygen molecules in blood are transported bound to haemoglobin, not freely dissolved in the plasma. So the only real way of increasing the amount of oxygen in the blood is to increase the oxygen saturation of the haemoglobin i.e. on average how many of its four oxygen binding sites are full. Haemoglobin has evolved as a protein to maximise the amount of oxygen it can take from normal air. Arterial blood (blood that has left the lungs) is over 95 per cent saturated with oxygen. This is why it is bright red as opposed to the dark blue/purple of the blood in your veins. There is not much room for improvement by breathing pure oxygen. Incidentally this is why it is hard to see how 'oxygen bars' or 'oxygen saturated' water drinks, both fashionable in the early years of this century, can be of any benefit for physical activities.

However, maybe during exercise the lungs don't manage to offload all their oxygen onto the haemoglobin? After all your blood is flowing very fast as your heart beats faster. In general during exercise at normal altitude, the haemoglobin in the arteries remains fully saturated, making increases in the oxygen content of the air largely redundant. However, in elite athletes the transit time in the lungs is so short that exercise-induced arterial haemoglobin desaturations are possible in some individuals; in these cases exercise performance can be enhanced by raising the air's oxygen content. The situation is very different when exercising at altitude. Here, as climbers on Mount Everest can attest to, the air pressure is so low that, even though the oxygen fraction is the same at sea level (21 per cent), the total number of molecules is much lower per breath. Arterial haemoglobin levels do desaturate and exercise becomes very difficult. Breathing pure oxygen clearly helps. For this reason it is very rare to have a major sports event at altitude—the distress of long-distance runners in the Mexico City Olympics of 1968 being a warning call that was heeded.

So how do you increase oxygen delivery to the muscle if not via changing the air you breathe? Increasing the heart's power (cardiac output) and/or increasing muscle blood flow will increase the total amount of oxygenated haemoglobin in the capillaries, the small blood vessels that surround the muscle cells. From there the oxygen transports via diffusion to the mitochondria inside those cells. Cardiac output, local muscle blood flow, muscle oxygen diffusion and mitochondrial content can all in principle affect maximal oxygen consumption. Biochemists and physiologists all have their own favourite mechanisms to champion. However, in general it is rare that a single factor has absolute control over a biological system. Instead there is a sensitivity to multiple effects.[1] The likelihood is that acute and chronic changes in all the parameters listed above all play some role in affecting $\dot{V}O_2$ max— changing any one can improve aerobic sports performance to some extent.

For one of these factors there is clear supporting evidence—the amount of oxygen in the blood. Studies starting in the 1970s showed that decreasing or increasing total haemoglobin content (via removal or reinfusion of red cells) can have essentially immediate effects on both $\dot{V}O_2$ max and running performance in the laboratory and in the field.[2, 3] Whether it is sensed in the brain or—more likely—directly limiting muscle function, the absolute amount of oxygen matters. So how does an athlete increase this? Not surprisingly there are both licit and illicit means to achieve this goal.

Physiological methods to improve blood oxygen content

Here the most prevalent method by far is altitude training. Even while focusing on the chemistry we shouldn't underestimate the psychological benefits (at least for UK athletes) of taking a break from the usual dull, wet environment to go to an exotic location. Especially when you are told that the trip will improve your performance. Nevertheless the performance benefit is not all in the mind. Living at altitude affects blood oxygen content directly. The mechanism is well understood.[4] The low oxygen pressure in the air is sensed by cellular oxygen sensors. This triggers the production of the now infamous peptide hormone erythropoietin (EPO) in the kidney. EPO regulates erythropoesis, the process controlling the development of new red blood cells. The more EPO you have, the more red blood cells you make. The body therefore compensates for the lower amount of oxygen bound to each haemoglobin molecule by just making more of them: crude, but effective. For an athlete who normally lives at low altitude, going to high altitude (above 2000 m) increases total haemoglobin content by about 1 per cent per week. Though this can be reduced if the athlete can tolerate going to significantly higher altitudes, in general the maximal benefits of altitude can require as long as eighty days. We can readily see the temptation to short-circuit the body's own physiological regulation.

Apart from the time required, another major problem with altitude training is that, by its very nature, it decreases the availability of oxygen to the tissue. So it is harder to train at altitude. It has always been a worry that any positive effects of increased altitude might be countered by a 'detraining' effect as the athletes have to reduce their training intensity to compensate for the lack of oxygen in the air. This has resulted in the 'live high, train low' method; athletes live and sleep at altitude for most of the day and night, but move to lower altitudes for their daily training. This cable car commuting has been shown to improve performance over the alternative, 'live high, train high', where the athletes live and sleep at altitude. The improved performance correlates with the change in the number of red blood cells.[5] Indeed the total red cell increase (8.5 per cent) was almost matched by the increase in $\dot{V}O_2$ max (6 per cent). This confirms that the increase in blood oxygen content is the key effect of altitude training.

There is an alternative to the 'live high, train low' method. This is to artificially change your inhaled oxygen pressure in the comfort of your own home environment. 'Portable altitude simulators' were used in the 1980s to force athletes to re-breathe their expired air, thus dropping the inspired oxygen. However, this use had been pre-dated by the systematic use of large-scale oxygen-deprivation chambers by East German, Scandinavian, and latterly, Australian teams. With the advent of small scale low oxygen tents dropping the price from hundreds of thousands of dollars to under ten thousand dollars, all elite athletes now have the ability to simulate high altitude at will. Although the theoretical ability of these facilities in raising red blood cell number is clear, performance enhancements are limited. It is likely that this is due to the length of time spent at low oxygen. At least 12–16 hours per day is suggested to be necessary to have real performance effects.[5] Overnight use of the tents alone is unlikely to be productive. The trip to the mountain resort wins both physiologically as well as psychologically.

Biochemical methods to improve blood oxygen content

There has been an active debate by anti-doping organisations about the ethics of using 'oxygen tents' (by which they mean low oxygen tents of course). Indeed they were recently banned from Olympic villages. However, what really annoys the authorities is adding a foreign compound into your body, especially via injection. The fact that these injections can instantaneously, and for no additional work, improve sporting performance places these techniques high on the World Anti-Doping Agency banned list. The chief villain is increasing your red cell content, not by hard work and altitude training, but by a process called blood doping. Just like a normal blood transfusion this involves simply lying down and having an infusion of red blood cells. However, unlike a therapeutic blood transfusion the recipient starts with an ample amount of red blood cells—the end result is therefore a super high concentration of blood cells not achievable by even the most rigorous altitude training regime. The cells can come from a matched donor, in which case the transfusion is called homologous. Alternatively the athlete can provide their own blood, in which case it is termed autologous. In the latter case the blood is first removed a long time (weeks-months) earlier and stored in the fridge. When enough time has elapsed for the athlete to return to a normal red cell count, the stored blood can then be re-infused. With either homologous or autologous blood the effect is the same: a clear and immediate increase in performance.[6]

Homologous blood doping requires a matched donor; autologous blood doping requires the acceptance of some training loss while the athlete recovers from the initial blood donation. A far easier, though slower acting, alternative arose when EPO became available as a drug. Like the majority of performance enhancing drugs EPO abuse has its roots in medical research. It is used as a treatment in situations where the body loses the ability to make its own EPO, such as the anaemia associated with end-stage kidney disease, or following cancer or HIV chemotherapy. Originally purified from human cadavers, like many biologically-derived medicines it is now made by bacteria using recombinant (Genetic Modification, GM) technology. GM technology is a safer method of production as it is guaranteed free from human pathogens. It also enables a much-improved rate of production. This led to increased availability of the drug—a fact not unnoticed by athletes and their coaches. Until out-of-competition drug tests became available recombinant EPO (rEPO) was widely used by athletes. This was most notable in the Tour de France, where police seized recombinant drugs from team support workers. The effects of rEPO are significant and readily out-perform the use of oxygen deprivation tents; as expected they seem to be mediated by the increase in red blood cell content and associated blood oxygen content.

EPO and blood doping work via enhancing the body's normal mechanisms of oxygen delivery. However, more exotic compounds are on the doping banned list. These are the so-called 'Blood Substitutes'—a class of compounds that have been in continual clinical development for the past twenty years. They have even found fame in the TV and movie industries with the recent penchant for vampire chic. Vampires need blood to survive; if they can drink artificial blood they can reduce their dependence on the human source and co-exist happily in normal communities. Outside the world of entertainment, blood substitutes are molecules designed to enhance the oxygen-carrying capacity of plasma (see Figure 9) rather than be a source of food. It is hoped that ultimately they could replace red blood cell transfusions in situations of rapid blood loss. Two main varieties of molecules have been tested—haemoglobin based oxygen carrier (HBOC) and perfluorocarbons (PFC).

HBOC are, not surprisingly, based on haemoglobin (Hb) itself. However, putting pure haemoglobin in plasma is not useful; outside the red blood cell the molecule falls apart and will quickly be cleared from the blood via the kidneys or attacked by the immune system. The biotechnological challenge is to modify haemoglobin such that it remains in the blood stream; this is usually achieved via a combination of some chemicals sticking the haemoglobin molecules together while others wrap around the surface as a kind of molecular disguise. The non-toxic molecules that enable the disguise are called polyethylene glycols. These have many uses in our synthetic world; few molecules find their way into blood substitutes, toothpaste, and laxatives.

A completely different approach is taken by PFCs. These unreactive molecules were originally synthesised to handle the corrosive uranium hexafluoride required to make the first atomic bomb. However, they have the additional property that oxygen readily dissolves in them. Consequently if PFCs are used to replace the water in plasma, blood oxygen solubility increases dramatically. This is a purely physical effect; no biological molecules are involved. This effect was most dramatically shown by Leland Clark. An American biochemist, Clark pioneered the development of heart-lung machines and invented the sensors that are now routinely used to measure oxygen consumption in athletes. However, his life's dream was to create artificial blood. In 1966 he showed that there was enough dissolved oxygen in a solution of PFC for a mouse to be able to swim quite happily completely submersed under 'water'.[7] The mice survived to live long and healthy lives.

Although they work by very different mechanisms, injections of either HBOC or PFC have been shown in clinical studies to increase the blood oxygen content and therefore enhance oxygen delivery. Hence their inclusion on the list of banned drugs. However, in these trials the patients are ill; they start from a situation where their oxygen delivery is compromised and treatment brings the oxygen delivery back to normal. It is less clear that blood substitutes can bring about enhanced oxygen delivery in people who *start* with normal blood oxygen levels, such as elite athletes.

In a healthy person haemoglobin is able to bind all the oxygen it needs in the lungs, even when the air is only twenty one per cent oxygen. PFCs are poor at delivering oxygen under these conditions. However, unlike haemoglobin, increasing the oxygen content of the air has a dramatic effect. The amount of oxygen dissolved in a PFC solution increases in a linear fashion with increased oxygen in the air. Therefore in hospitals PFCs are always used with the patients breathing 100 per cent oxygen gas. But this poses obvious problems for athletes. It seems unlikely that carrying an oxygen cylinder around with you is going to go unnoticed. In the absence of inhaled oxygen gas we are looking at very small increases in oxygen in the blood following PFC doping. In the case of HBOC the situation is rosier for the dopers and there are human studies suggesting the possibility of enhanced exercise effects.[8] However, these results did not include performance tests. Increasing blood oxygen content outside the normal regulatory control of the red cell may be deleterious and there is no substitute for proper field tests.

Even if they were shown to be performance enhancing, HBOC have a number of problems with toxicity. Broadly speaking HBOC are dangerous because they can kill good radicals and make bad radicals.[9] The good radical is nitric oxide. As we saw in the last chapter, this gas is a key regulator of blood flow, blood volume and blood pressure. Haemoglobin can destroy nitric oxide. Normally the red blood cell acts as a protective barrier, keeping nitric oxide and haemoglobin apart. However, when an HBOC is added to plasma there is no such defence and the nitric oxide is rapidly removed. This causes a decrease in blood flow, not a good side-effect for a molecule designed for the emergency enhancement of oxygen delivery around the body.

But there is worse to come. HBOC can also make free radicals themselves. Free radicals are reactive molecules that are used by the body to catalyse chemical reactions or for cell signalling. But in excess, or in the wrong place, they can be toxic. A free radical is a species with an unpaired electron. Electrons prefer to live in pairs. So free radicals desperately crave electrons. Removing an electron is chemically called

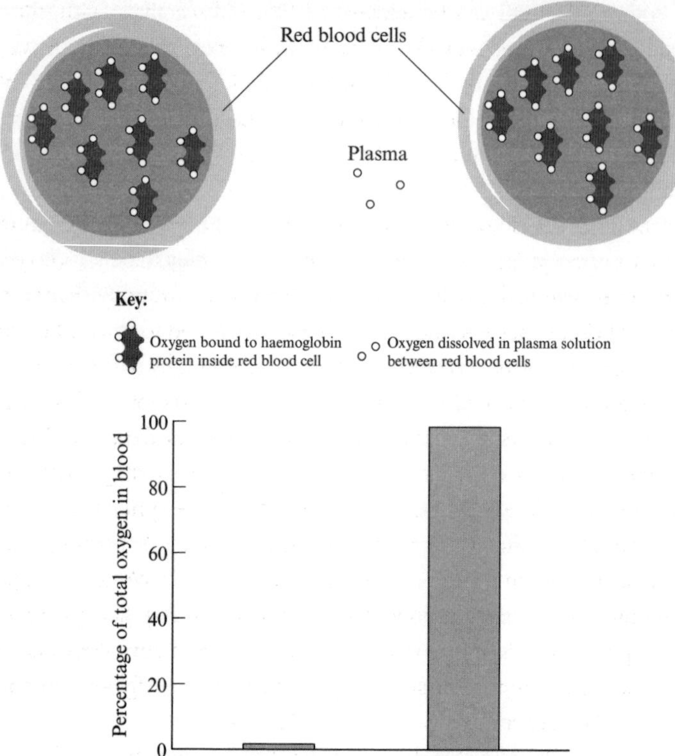

Figure 9 Comparison of the amount of oxygen in plasma and red blood cells

The bar chart illustrates the huge increase in the oxygen content of blood caused by the four molecules of oxygen that bind to each haemoglobin protein molecule.

an oxidation (derived from oxygen itself which has unpaired electrons and therefore can oxidize other chemicals). So free radicals tear electrons away from cellular molecules in their desire to pair up. This is why the body has antioxidant defences, such as the vitamin C and vitamin E we take in our food (or more lazily nowadays in pills and or soft drinks).

Although HBOC can be engineered to avoid reactivity with nitric oxide, the other free radical toxicity problems have proved less tractable. As a scientist working in this field would I take one of the current generation of these blood substitutes if my life were in danger? Absolutely without question. Would I think I was taking a risk of side-effects? Yes.

The situation is no rosier with PFCs. The haemoglobin molecule in HBOCs is recognised by the body and not attacked by the immune system. Then, like all drugs, it is cleared gradually from the body. However, many PFCs are inert—remember they were designed to be resistant to bombardment by the products of the radioactive decay of uranium. So they can stay in the body for a very long time—years for the first generation of products. Also, as a genuine foreign chemical there is always a concern that the body's immune response will reject PFC particles as 'foreign'. Therefore despite over twenty years of development and clinical trials HBOC and PFC's are not currently considered safe enough for patient approval in the USA.[10] A similar view is held in Western Europe although Hemopure, a product based on modified cow haemoglobin, was approved for use in South Africa and Russia in 2011. Perftoran, a PFC, was used by the Russian army in their war in Afghanistan and approved for general clinical use in Russia in 1997.

A potentially safer alterative to artificial blood substitutes is the use of embryonic stem cells to make synthetic red blood cells in the laboratory.[11] Theoretically these cells should be identical to naturally occurring red blood cells. Three groups around the world are leading the way in this field. In Scotland and France the work is largely funded by universities and public agencies. In the USA, however, due to the political constraints on stem cell research in the early part of the twenty-first century, research largely occurs in the private sector. All groups have managed to make red blood cells. The difficulty is making enough for a transfusion.

Whether making completely artificial blood or growing stem cells the major problem is one of industrial biotechnology. Blood is like no other drug. You just have to make so much of the stuff. Previous attempts to make artificial blood from bacterial cells floundered on the scale of the process involved. Fermentors the size of those in breweries are needed to grow enough bacteria. In the seminal 1970s movie *Chinatown*, the crime and corruption in Los Angeles ultimately involves a struggle to control the water supply necessary to expand a city of many millions in an arid environment. The same principle holds in manufacturing blood. You need to follow the water (as well as the money). Two early pioneers in the field, Somanetics (USA) and Delta (UK) were based in Boulder on the Colorado River and Nottingham, on the River Trent respectively. I heard representatives quip that these facilities were fortuitously positioned, as, if successful, they would need to harvest a high percentage of the water from these rivers to make enough artificial blood for worldwide use.

The ideal stem-cell-derived blood would be available for universal donation—effectively a super type 'O' with no markers for immune rejection by the recipient. For an athlete this would be problematic, as such a red blood cell would be immediately detected as 'foreign'. In principle it would be possible to clone a red blood cell from an athlete and use that as an undetectable supply of blood for doping. But as the present technology using the more adaptable embryonic stem cells is 5–10 years away, it is hard to see the production of red blood cells from adult sources being a possibility any time soon. The idea of an athlete sending off a drop of his blood to a lab and waiting down the line for a courier to deliver him a bottle of his own red cells is not something we will be faced with in the foreseeable future.

Making the most of oxygen

A: GETTING OXYGEN OFF HAEMOGLOBIN

Increasing the oxygen content of the blood is not enough to increase the oxygen content of the mitochondria. The oxygen needs to get from source (haemoglobin) to cell to mitochondria. The first part of this process is offloading oxygen from haemoglobin. This has attracted the interest of sports scientists recently. Haemoglobin has evolved to pick up oxygen from the lungs and deliver as much as it can to the cells. In healthy people, as described previously, the molecule can pick up 100 per cent of the oxygen it needs from the lungs. By the time it has passed through the muscle of an elite athlete less than 15 per cent of the oxygen remains.[12] Is there a way to offload even more of this oxygen from haemoglobin?

The amount of oxygen that can be delivered is a function of the oxygen binding properties of the haemoglobin molecule. Haemgolobin needs to bind oxygen tightly enough so that it can capture it all from the oxygen-rich environment of the lungs, but not so tightly that it doesn't give it up to the oxygen-poor environment of the muscle. This is a delicate balancing act. There is one compound on the doping banned list that affects the oxygen binding properties of haemoglobin. It is called efaproxiral.

Efaproxiral has its roots in medical use. It was designed to deliver oxygen to tumors. Radiotherapy works by creating oxygen-derived free radicals. This radiation damage requires oxygen. So the more oxygen that is present in a tumour the more effective radiation therapy is in creating toxic oxygen radicals. Efaproxiral is also known as RSR13 (right shifting reagent 13). The 'right shift' refers to a graph of oxygen binding to haemoglobin. Shifting this curve to the right makes it easier for haemoglobin to offload oxygen at the low oxygen content in a cancer cell. The idea being that if you can get more oxygen to the cancer cell radiotherapy will be more efficient.

If RSR13 allows haemoglobin to release oxygen more readily in cancer cells, it could also do so for muscle fibres. Can this be utilized by elite athletes? In animal studies RSR13 has been shown to increase $\dot{V}O_2$ max across an electrically stimulated muscle. However, what you gain at one end you can lose at the other. The right shift of the binding curve makes it easier to get oxygen off haemoglobin in the muscle but harder to get it on in the first place in the lungs. If the haemoglobin molecule binds oxygen more weakly it may not be able to grab as much oxygen from the lungs. In the animal studies described above, and also with cancer patients, the haemoglobin oxygen saturation in the lungs was kept artificially high by breathing 100 per cent oxygen.[13] It is unclear in a human athlete breathing normal air whether the gain in the oxygen delivered to the muscle would offset the loss in the oxygen picked up from the lungs. Nevertheless the theoretical benefits are enough for it to be banned in all sports.

B: GETTING OXYGEN FROM THE RED CELL TO THE MUSCLE

Once off the haemoglobin, oxygen still has to get to the muscle mitochondria. Oxygen diffuses relatively slowly in watery solutions such as plasma. It could be advantageous to increase this diffusion. One way might be by having a molecule that could rapidly bind and release oxygen. To visualize how this could work imagine the classic case—beloved of old black-and-white films—when a line of stars and extras toil to put out a fire by making a human chain to pass the buckets from water pump to fire. There is a molecule that already performs a chemical version of this chain inside muscle cells. It is myoglobin, a muscle protein closely related to haemoglobin (myo- coming from the Greek word for muscle). Myoglobin is present in high concentration in all muscle and heart cells. It is faster for oxygen to diffuse via binding and rebinding to myoglobin molecules than to move through the cell solution alone.

Could the same thing be possible in plasma? It is clear that oxygen diffusion in plasma is limited by its poor diffusion through the water solution. Putting a rapidly diffusing oxygen binding molecule into this solution will enhance oxygen transport through the plasma. In this case, like myoglobin, the protein would not be acting primarily to increase the blood oxygen content, but instead to increase the rate of oxygen transport to the cell. Indeed the beneficial oxygen transport properties of one HBOC undergoing current clinical trials, Hemospan®, is attributed, at least in part, to possession of exactly these properties.[14] Whether any effect of low dose HBOCs on improving plasma oxygen transport in healthy elite athletes would improve performance depends, of course, on how much this diffusivity limits oxygen delivery under these conditions; the sporting environment is very different to those in a clinical trial of blood substitutes. Nevertheless this is another potential source of oxygen limitation in aerobic exercise that could be overcome by pharmacological manipulations.

Evidence in the field? What techniques are used to enhance oxygen delivery to the muscle?

Most of the techniques described have indeed been used by athletes in competition. On the legal front altitude training is a major training component of most elite athletes involved in long term endurance events. Low oxygen tents for individuals (or more rarely rooms for groups) are less common, but still occur. Of the current banned methods blood doping came to prominence in the 1970s and 1980s with both Finnish and Italian athletes and members of the 1984 US Olympic cycling team later admitting to its use; blood doping was not formally prohibited at the time, although the revelations were treated with considerable disquiet. For a while blood doping took a back seat to EPO, but more sensitive testing has revealed that it has not gone away. Historically doping with homologous blood has been preferred given that—as opposed to donating your own blood—there is no loss in training following the blood loss. There are persistent rumours of 'ringers' sent to games merely to provide blood for the elite athletes; there are even suggestions that weaker members of teams have been specifically selected for elite competitions, not for their performance alone, but because they can double up as compatible blood donors for the 'team leader'.

Equipment consistent with the use of blood transfusion was confiscated from the Austrian team at the winter Olympics of 2002 and 2006. In 2002 the Austrian ski federation's defence was that that the needles, tubes and transfusion bags were needed so that they could withdraw blood and then expose it to a magnetic field and ultraviolet radiation prior to reinjecting into the body.[15] This, it was claimed, was an effective cold prevention remedy in common use at many spas. The authorities were not impressed; the IOC banned the athlete's coach, the Norwegian Walter Mayer, from attending the 2006 and 2010 Olympics. The 2006 story was worthy of a Hollywood movie, or perhaps more appropriately a reality TV police show. The Austrian team were based in the quiet Italian mountain towns of San Sincario and Pragelato. Mayer was spotted, despite his ban. During a subsequent police raid on the team's accommodation bags of used syringes were seen being thrown out of the window. Some athletes and coaches promptly fled. Mayer was caught when he crashed into a police roadblock in the Italian Alps. In 2007 six skiers were given lifetime bans from competing at the Olympic Games; in 2011 Mayer was sentenced to a fifteen month jail sentence for supplying banned substances.

There have been other high profile cases. Tyler Hamilton is a US cyclist whose career seemed beset by bad luck; he broke two vertebrae in a ski jumping accident prior to taking up cycling and fractured his shoulder whilst finishing second in the Giro d'Italia. When he first joined the Tour de France his role was merely to support his team leader—Lance Armstrong—on the mountain stages. Even his greatest success on the Tour was achieved in the face of adversity. In 2003 he cracked a collar bone on the first stage. But he cycled through the pain to win stage sixteen with a 142 km solo breakaway, finishing fourth overall in the final tour. Later he founded the Tyler Hamilton Foundation for multiple sclerosis research that raised funding by taking cyclists of all abilities over some of the most famous mountain passes in the world from Alpe d'Huez to Col St Bernard.

But there is another side to Tyler Hamilton. He was found on multiple occasions to have more than just his own population of red blood cells in his body. The story starts in April 2004. His blood was found to have a high ratio of old to young red blood cells indicative of a recent infusion of red blood cells or EPO doping. The normal value is 90. The UCI (Union Cycliste Internationale) suspends someone if the score exceeds 133. Hamilton's score was 132.9. This sample also showed someone else's blood was present, but this test was not then considered validated for official use. Then in August, Hamilton had his greatest individual success, winning the gold medal in the 2004 Olympic men's individual cycling time trial. By this time there was a robust validated test in place for blood doping. Hamilton was again found to have had foreign, homologous blood in the first blood sample he donated. But his backup sample was frozen and could not be tested. So he kept the gold medal despite a later protest from the second-placed Russian athlete.

However, in September the authorities finally got their man. Following a Spanish road race, both blood samples were positive for doping and Hamilton was banned for two years. Throughout the ban period, Hamilton vigorously protested his innocence. His lawyer even suggested Hamilton was a 'chimera' i.e. he had two red blood cell lines, the result of a non-identical twin merging with his cells at a very early stage of development. This would then be responsible for the apparently foreign blood cells. Even Hamilton disowned his lawyer's fanciful theory, but could not provide other evidence to convince the authorities to repeal his ban. He returned to professional cycling in 2006, still strongly protesting his innocence. Three years later, he was banned for eight years for taking DHEA (Dehydroepiandrosterone), a steroid precursor that was part of a supplement he took to counter his depression. He retired from cycling.

Nevertheless it is unfair to pick on one individual, however high profile. When it comes to cycling, blood doping is endemic. The most famous unsolved mystery is the 99 bags of blood plasma seized as part of Operación Puerto, a high profile 2006 Spanish police investigation into the blood doping allegedly masterminded by Dr Eufemiano Fuentes. Although the police investigation stalled in 2007 with no criminal charges instigated, over fifty riders were implicated in this scandal; many were prevented from starting the Tour de France in 2006, including potential winners in Jan Ullrich and Ivan Basso. The key to unravelling the case was deciphering the code names used for athletes on their blood bags. Not all were as easy as Ivan Basso, who used his dog's name Birillo. Basso has since admitted he intended to dope himself by infusing this blood, but denies going through with this; nevertheless he received a two-year ban from the Italian cycling federation for attempting to use a banned substance or method and 'possession of banned substances and methods'. Ullrich on the other hand denies any involvement with Fuentes. An insight into the minds of those involved in this process can be seen in an interview with one of the few who admitted to being involved—the German cyclist Jörg Jaksche in a 2007 interview with *Der Spiegel*:

> Yes, I did dope, but I never overdid it. I never took artificial haemoglobin or stuff like that, where you can get an allergic shock. And you calm yourself by saying that a guy who does bodybuilding takes 16,000 units of growth hormone a day, and I only took 800 units once in a while for regeneration. Then you think: Well, it's not that much after all.

EPO injections avoid the storage problems of blood transfusions. Though it is likely to have been taken much earlier, the use of EPO came to prominence in the late 1990s. A key member of any elite cycle team is the *soigneur*—basically a cross between a gopher, fixer and a therapist who is responsible for feeding, clothing, and even massaging riders. Willy Voet was the Belgian-born *soigneur* for the Festina team. For the full drama of these events it is worth reading Voet's 2001 confessional book *Breaking the Chain* (see Chapter 1 ref. 2). It opens with Voet in an official Tour de France car being stopped by customs officers on the Calais road en route to the start of the tour. Voet is worried about being caught with the 'Belgian mix' he is carrying—a potent mix of amphetamines, caffeine, cocaine, heroin, painkillers and corticosteroids used by the cyclists. His worry is compounded by the knowledge that he had taken some Belgian mix himself to keep alert on the long car journey.

In fact the officials were more concerned with the 234 doses of EPO in a refrigerated bag in the boot of Voet's car. This discovery kick-started the investigations into doping in elite cycling that have continued pretty much unabated since. Over fifty professional cyclists have tested positive for, or admitted, taking EPO to enhance their performance. As most of the admissions come safely after retirement, and not all cyclists are tested for EPO at any specific race, it is safe to say that this number of fifty is a large underestimate. Nor are cyclists the lone culprits. The Chinese authorities withdrew a number of their rowers from the Sydney 2000 Olympics after positive EPO tests prior to the games. Track athletes, starting with a 3000 m steeplechaser in 2002 (Brahim Boulami) have also fallen foul of EPO testing. Boulami has run the fastest 3,000 m steeplechase of all time; despite denying he had used drugs, he was banned for two years by the IAAF for EPO use and his world record was overturned. Two other runners Zheng Yongji and Li Huiquan, were expelled from the fifth Chinese City Games in 2003 and banned by the Chinese Olympic Committee for three years.

Just as with designer drugs and steroids, synthetic derivatives of EPO have not escaped the attention of athletes. Due to the clinical importance of EPO, there is significant interest in the biotechnology and healthcare initiatives to create modified forms of EPO that could be either cheaper to make or more potent in effect. EPO consists of a protein linked to some sugar molecules. This mixture of protein and carbohydrate is termed a glycoprotein. It is possible to make a version with altered carbohydrates that can last longer in the body. Darbepoetin is just such a synthetic molecule with an extended half-life in the body and hence an increased potency. It is used to treat anaemia associated with chronic renal insufficiency and chemotherapy. Thinking the use of a modified drug made them untestable, many athletes switched from EPO to darbepoetin. Alas for them the drug testers had worked out a test in secret. In 2002 Russian (Olga Danilova and Larissa Lazutina) and Spanish (Johann Mühlegg) cross-country skiers were banned for two years by the International Ski Federation whilst Russian (Faat Zakirov) and Italian (Roberto Sgambelluri) cyclists received one year and six month bans from the cycling authorities (UCI and FCI).

Recent modifications of EPO contain artificial mixtures of chemical polymers and biological amino acids. The first of these to come to market incorporates polyethylene glycol. Termed Continuous Erythropoiesis Receptor Activator (CERA) this molecule is longer lasting than EPO and was just about to be licensed to help patients with kidney problems in 2008. Even before being given to the first patient, the compound found its way into the doper's armoury. Four riders in the 2008 Tour de France, including the original third place finisher, Austrian Bernhard Kohl, were caught by a CERA test introduced by the French anti-doping agency. Armed with the new test, all the blood samples taken in the Beijing 2008 Olympic Games were subsequently re-tested for CERA doping. Five athletes were caught retrospectively. One was Italy's silver medal winning cyclist Davide Rebellin who denied wrongdoing, but had his medal taken away by the IOC following confirmation of the test results. Another was the German cyclist Stefan Schumacher who had the dubious honour of testing positive in both the Olympics and the preceding Tour de France. Although he appealed to the Court of Arbitration for Sport, they upheld the two year ban imposed by the UCI. The other three were in track and field events; the most notable of these was the double world champion Rashid Ramzi, who came first in the blue riband event—the men's 1500 m. Like many elite African athletes Ramzi had changed his nationality to that of one of the oil rich Gulf States—in this case from Morocco to Bahrain in 2002. Briefly he held the honour of being the country's first Olympic gold medal winner, with his victory achieving royal acclaim. In the words of King Hamad Bin Eisa Al Khalifa 'This outstanding victory consecrates Bahrain's international sports status'. In 2010 Ramzi was banned for two years by the Bahrain Athletics Association. *Caveat emptor.*

Following the clinical success of darbepoetin and CERA, new compounds are on the drawing board. Some have little in common with EPO except an ability to bind and activate its receptor. The first such completely synthetic mimic of EPO is called Hematide and is undergoing final phase III clinical trials for treatment of anaemia starting in 2010. The future for medicine is bright for EPO-type drugs and there is no doubt that dopers and drug testers will be fighting this battle for the foreseeable future.

The more exotic oxygen delivery therapies have proved less attractive to dopers. There is no evidence that the haemoglobin modifier RSR13 has been used in sport. Although indirect evidence suggests that some athletes may have been attempting to use blood substitutes, given the toxicity of both perfluorocarbon and haemoglobin-based blood substitutes, it is perhaps just as well that these appear to have been isolated incidents. The Swiss cyclist, Mario Gianetti, was investigated for taking perfluorocarbons—whether knowingly or unknowingly—when he was admitted into hospital in 1997, but no charges were made. Perfluorocarbons also featured in the Festina affair after the cycling team's offices were raided following the arrest of Willy Voet in 1998.

But my favourite story is about the cyclist Dario Frigo at the 2001 Giro d'Italia. Two vials of the blood substitute, Hemassist™, were confiscated in raids by police and he was ejected from the race and sacked by his team. Frigo admitted that he had bought the haemoglobin-based blood substitute on the internet, but never used them. Whilst he could have been forgiven for not knowing that this compound is best kept in a frozen state until immediately prior to use, the fact that it was colourless might have been a give away. Haemoglobin is after all what gives blood its red colour. The solution, which Frigo never took, was later shown to be a harmless salt solution.

The future

What does the future hold for oxygen doping? It is clear that these methods can provide significant performance enhancements. Therefore the history of sport tells us that there will be an ongoing struggle between athletes trying to bend/break the rules and drug testers trying to catch them. The appearance of designer steroids such as THG not originating in clinical medicine has opened up the whole doping field to new challenges. Oxygen doping is not likely to escape. Apart from the obvious targets such as more EPO derivatives there are likely to be a panoply of compounds that will affect red cell development of which we are as yet unaware.

EPO is just part of a signalling pathway that controls red blood production when the body responds to oxygen.[17] The body has a range of proteins that sense the low oxygen in the first place, the best characterised being a molecule called HIF (hypoxia inducible factor). HIF is activated by a fall in oxygen. It then triggers a range of events in the body, one of which is EPO production and the consequent red cell increase. So taking a molecule that enhances HIF function should have the same beneficial effect as doping with EPO. HIF is unstable and is designed to break down naturally and turn off its activation signal. But doctors are keen to create molecules that can stabilise HIF to prevent oxygen deprivation in organs stored for transplantation.

Acting to enhance the HIF protein could enhance the function of transplanted organs such as kidneys; there is also military interest as HIF activation could be a quick way to prepare troops for high altitude operations. Whether the same HIF activators could replace EPO as a sports doping agent is not clear. However, I wouldn't bet against it; with the perfect storm of medical and military interests coinciding, there should be no shortage of funds to advance this area. One concern for the general public, but maybe not elite athletes, is that HIF is very active in tumours; it helps them survive in the low oxygen environment.[18] Activating HIF may therefore help any nascent tumours to develop. This might stop a drug being developed clinically, but might be considered worth the risk for elite athletes.

HIF has a much more widespread function than merely increasing the levels of EPO. For example it can also increase the growth of blood vessels. HIF activation is how tumours grow blood vessels around them to deliver their oxygen. It is possible that long term use of a HIF activator could allow an athlete to remodel his blood supply and deliver more oxygen to his tissue. This anatomical advantage could outlast even the removal of the drug. This production of new blood vessels triggered by HIF is mediated by a molecule called vascular endothelium growth factor (VEGF). VEGF itself is therefore another potential target for improving blood supply to muscle tissues.

As you delve into this field it quickly becomes apparent that the drugs that are currently banned are only a subset of a much greater variety of possible compounds acting on the same, or even completely different, biochemical pathways. This is seen even more clearly when we turn to the next branch of sporting aids—those that target strength not endurance.

5 Muscling Up

'Are you tired of sand being kicked in your face?
I promise you new muscles in days!'
Charles Atlas

We need oxygen even to go for a gentle stroll in the park. But we don't need it to throw a punch hard enough to knock someone out. A short sharp show of strength is all about the power you can generate from your muscles. For this the fuel stored in each muscle cell is more than adequate. What you really need to land a forceful blow is as much muscle as possible. In this case size is indeed what matters; and when it comes to size muscle cells are amongst the giants of the body. Most people's perceptions of cells are that they are small entities only visible under a microscope. However, this is not always the case. Neurones are cells that carry fast messages in the body such as the electrical signals in the brain. Some of these neurones need to signal over very large distances. The longest cell in the body is the 1.5 m long sensory neurone needed to send an electrical signal from the toe to the base of the spine. Whilst not quite reaching these limits, muscle cells, or muscle fibres as they are usually called, can also be very long. The sartorial muscle for example—so named as it is the one a tailor uses to determine an inside leg measurement for a new pair of trousers—runs the length of the thigh and can contain fibres as long as 60 cm. A single muscle can contain over a million of these long fibres—optimised for coordinated contraction and relaxation to create forces for rotation, extension or flexion.

How to get stronger? Growth versus division

An elite athlete keen to improve their strength will obviously want to enhance the performance of these fibres. One can envision two mechanisms to achieve this. Increasing the size of an individual cell is called hypertrophy (after the Greek *huper* meaning overmuch and *trophikos* meaning nourishment). Alternatively increasing the number of cells is called hyperplasia (after the Greek *plasis* meaning moulding). What route is chosen is forced in part by the unusually long cell length of a fibre.

Long cells face special problems. A diagram of a cell is usually illustrated with a single, dark nucleus that contains all its genetic material (DNA). This genetic material is far from passive—every moment the DNA is working to make a molecule called RNA, which then triggers the synthesis of new proteins to enable the minute-by-minute function of the cell. But unlike an electrical signal which moves very fast, molecules like RNA have to physically move through cells—to visualise this imagine injecting some ink into a beaker of water and seeing how it slowly diffuses. To counter this diffusion problem muscle fibres have not one, but tens of thousands of nuclei. This allows proteins to be synthesised along the entire length of the muscle. However, these nuclei cannot divide. This means that in a mature adult it is generally considered that a muscle cell also cannot reproduce—you are stuck with the number of muscle cells you are born with. Muscle fibres have given up sex for power.

There are some exceptions. In fish, new muscle cells are made throughout development. Even in humans extreme exercise has the potential to induce muscle fibre splitting; one cell becomes two. While this is generally viewed as a form of damage, rather than productive cell division, the exercising of mice and rats has been shown to cause a genuine twenty per cent increase in muscle cell number.[1] Even more profound effects are seen in birds; increases by as much as eighty per cent in muscle cell content can be measured in the Japanese quail after muscle stretching.[1]

How can we check whether exercise can increase the number of muscle cells in adult humans? The problem is that unlike blood or urine samples, people are reluctant to donate muscle samples for testing; a muscle biopsy is painful and, according to the prevailing theory, new cells will not grown back to replace the old ones. So there is no chance of an elite athlete donating part of their muscles for scientific research. Even in volunteers only small samples the size of orange pips are used—the resulting human evidence base therefore lags behind animal work.

It is also questionable how relevant the animal work is to human muscle building. In the case of the bird study described above the wings were 'stretched'. This is equivalent to having 35 per cent of the total body weight stuck on one wing for twenty-four hours; the other wing being left as a control. This is not the equivalent of the average training programme in the gym. Nevertheless it does illustrate that the inability of a muscle cell to divide may not be as complete a barrier to the development of increased muscle strength as we once thought. If laboratory animals can show an increase in weight via an increase in muscle cell number maybe, given the right training regime—or pharmaceutical enhancement—we can do the same.

The training effect

Current weight training programmes, whether drug-enhanced or not, primarily work by increasing muscle fibre size, not number—hypertrophy not hyperplasy. Most biological cells consist of a controlling nucleus surrounded by a cytoplasm in which molecules such as proteins, fats and carbohydrates happily float around. However, muscle cells have in addition an array of filaments predominantly made up of just two proteins, actin and myosin. Using the energy from ATP these filaments slide along each other contracting and expanding the muscle fibre. The combination of millions of filaments in thousands of fibre cells simultaneously contracting allows the muscle to do gross anatomical work, like lifting weights or sprinting out of starting blocks.

Strength-based training programmes cause changes in body shape that we are all familiar with; underlying this increased musculature is an increase in the size of individual muscle cells, not their total number.[2] This muscle cell hypertrophy can be achieved in two different ways. The first is by increasing the contents of the cell constituents that are responsible for energy metabolism. These comprise the cellular and mitochondrial enzymes that break down glucose into ATP. To encompass these extra molecules the cell volume expands. The cell is now 'primed' to use more energy to work harder. However, this is no good in the absence of new actin and myosin filaments; these intracellular protein scaffolds are necessary to allow the newly enlarged cell to make use of all its extra energy to do work. So an increase in myosin and actin must follow if you want the enlarged cell to force a more powerful muscle contraction.

You might assume that athletes would seek training regimes and/or drugs that optimise the cellular energy contents and the myosin actin filaments—thus giving the muscle cell the tools for contraction and the energy to power those tools. For most sports this is true. Weight lifters and sprinters optimise their strength by lifting weights close to their maximum for short periods of time. However, professional bodybuilders tend to lift easier weights for longer periods. This requires more use of aerobic energy resources and leads to growth dominated by the ATP-generating systems, not the actin and myosin filaments. The result is that the body builder's cells expand and they look good, but they don't have the power to match. If you just exercise to look good then it is likely the beach bully will still be able to kick sand in your face with impunity.

Training at the cellular level: the role of biological signaling

A compound that enhances training must work at the microscopic level inside the cell. In the case of muscle growth key messages are sent by specific molecules to inform the cell to initiate the complex range of processes required to grow. Just like any construction project this involves laying in materials, project managing the integration of resources and the build itself. The blueprint for this growth process is already in the muscle cell. To increase muscle mass what an athlete needs to do is initiate the start signal. Pushing this start button harder and/or more often is the key to optimising strength training.

Biological signalling works on a precise 'lock and key' mechanism. A target receptor provides the lock and requires a precisely matched molecular key to open it. The key is provided by hormones—small biological molecules that are made in one part of the body and affect a process in another part. Typically they are released in one cell and pass through the bloodstream where they target a cell in a different organ. A typical example is insulin. Special cells in the pancreas sense the level of glucose sugar in the blood. If blood sugar levels increase, insulin is released and travels to other cells—for example muscle fibres—to make sure the excess blood sugar is taken up from the bloodstream. This at once rebalances the system and at the same time delivers sugar to the cells that can make best use of it.

A hormonal signal is akin to the signals on a train track. It can tell the body's metabolism to start, stop, go in one direction, go in another or even reverse direction altogether. What is the secret to this flexibility? Just as we saw with ATP and energy metabolism, the predominant mechanism relies on phosphate chemistry. Protein enzymes control biological reactions. Specific amino acids on these proteins can bind phosphate, a process called phosphorylation. This phosphate alters the reactivity of these enzymes making them able to catalyse their chemistry faster or slower.

Two factors change this simple model into the flexible system that controls our biology. The first is that the phosphate bond can be readily synthesised or destroyed. Just as we saw with ATP this phosphate bond is neither too strong nor too weak. This is key. It means that any change in activity is not a one-way street—a signal can be turned off as well as on. The second factor is that the phosphate addition and removal is catalysed by enzymes *that are themselves subject to control by phosphorylation and dephosphorylation.* This offers opportunities for massive amplification of signalling. A single change can readily be amplified many times, in what is known as a signalling cascade. Figure 10 illustrates this process.

The two signalling pathways key for muscle growth are called mTOR and myostatin; both have such protein phosphorylation at their core.[3] One pathway signals for an increase in protein synthesis and the other signals a decrease. mTOR is the 'GO' pathway and myostatin the 'STOP' pathway. The pathways are named after their key phosphorylated proteins. Myostatin quite sensibly is derived from the Greek words 'myo' meaning muscle and 'statos' meaning stop. Unfortunately the m in mTOR does not stand for muscle. Instead mTOR is an acronym for 'mammalian Target of Rapamycin'—a bit of a mouthful.

mTOR is a biochemical pathway that turns on protein synthesis in a number of different cells, hence the lack of a muscle-specific name. Its name derives from the fact that an antibiotic rapamycin targets this key protein regulator to inhibit cell growth. Rapamycin is produced by a fungus that was first discovered growing on Rapa Nui (Easter Island) in the Pacific Ocean. It is used in medicine to prevent the immune response attacking donated organs following transplantation. By inhibiting the mTOR pathway it can also stop cells synthesising new proteins and therefore growing. This gives it anti-cancer properties as it can stop cancer cells dividing. I still await the quiz question that asks for the link between bodybuilders, a cure for cancer and the stone statues of Easter Island.

Altering these pathways with drugs—turning on the GO *mTOR* or turning off the STOP *myostatin*—has the potential to build up muscle mass and hence power. So it is worth us taking a detailed look at these pathways—something the dopers have already done of course.

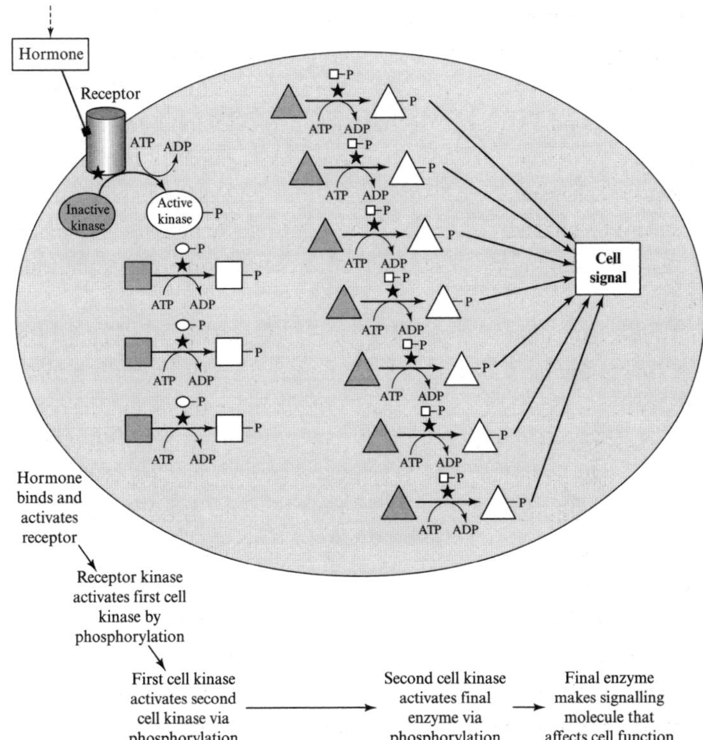

Figure 10 Amplification of hormone signalling by a phosphorylation cascade

The stars indicate catalysis (speeding up of a reaction) by a kinase enzyme adding a phosphate group and converting the enzyme from an inactive to an active state. A hormone arrives from the blood and binds to its highly specific receptor on the surface of the cell; this event can then activate many enzymes to cause an amplification of the signal. Turning the signal off requires the opposites—a dephosphorylation cascade (this uses a separate set of enzymes that remove phosphates called phosphatases).

mTOR: the start button

mTOR integrates outputs from a number of pathways; the resultant signal if large enough (GO!) can tell the body to convert amino acids into proteins—the necessary prerequisite for growing muscle cells. A useful analogy is the old coin drop machines at seaside arcades. Different people push coins in of different shapes and sizes. A tantalising stack of coins sits precariously over the exit hole. Eventually, if enough coins arrive at the right place at the right time, the stack topples over and is released into the expectant hand of the player. The process then repeats.... Signalling pathways work on a similar analogue integration principle.

So what are these inputs? Figure 11 illustrates the mTOR pathway. Inputs can be divided between internal control mechanisms and external signalling hormones. The internal control mechanisms are the body's way of 'deciding' whether there is a physiological need for muscle growth and, if so, whether there are the biochemical resources to fulfil this need. The mTOR pathway senses, and is activated by, mechanical stress, high amino acid levels and energy. Basically mTOR is activated if the muscles are worked hard *and* there are plentiful building blocks for new protein synthesis (amino acids) and the energy (ATP levels) to drive this synthesis. The combination of a hard muscle work out with a healthy food energy input are exactly the conditions that athletes aspire to during weight training. mTOR is one of the biochemical rationales underpinning weight training programmes.

In principle these physical and biochemical signals must ultimately act via a change in the activity of enzymes inside the muscle cell. These intracellular enzymes are therefore potential targets for doping agents. However, it is difficult to design a new drug to target specifically the inside of a muscle cell. Hormones, on the other hand, have evolved to do exactly this i.e. be produced in one location and move to another. So the specific human hormones that can enter muscle cells and activate mTOR are attractive to the doper. These can be synthesised artificially outside the body and given to the athlete to enhance their natural levels.

They include insulin and a similar molecule called insulin-like growth factor 1 (IGF-1). IGF-1 can be modified in cells in the laboratory to create a new molecule called IGF-1Eb, which may be even better at activating mTOR. IGF-1Eb has therefore acquired the catchier moniker of mechano growth factor (MGF). It is not clear whether MGF is ever made naturally in the body. But it is readily synthesised in the laboratory. All of these external agents—insulin, IGF-1 and MGF—have the possibility to increase muscle growth. All three molecules, not surprisingly, feature high on the list of banned substances in sport.

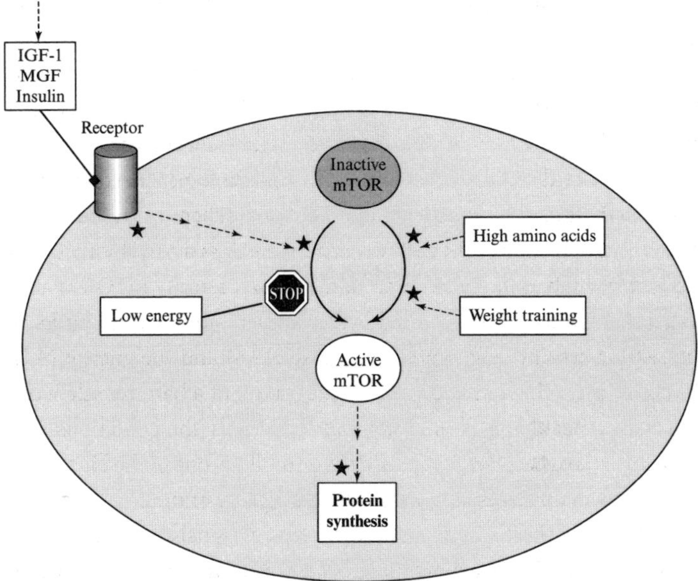

Figure 11 The mTOR 'GO' pathway for turning on protein synthesis

A simplified picture of events leading to an increase in protein synthesis following a hormonal signal. The stars indicate catalysis (speeding up of a reaction). The stop sign indicates a fail-safe mechanism where energy is not used to synthesise proteins when energy stores are low. The dotted lines indicate a complex series of signalling/phosphorylation cascades as per Figure 10. Note that mTOR activation by weight training and food (amino acids) does not necessarily require a hormonal signal.

However, there is a problem with targeting the mTOR pathway. It is a general pathway the body uses to increase protein synthesis in all cells, not just the muscles. The addition of hormone activators may therefore have undesirable consequences. Inhibiting mTOR with rapamycin has anti-cancer properties by stopping cancer cell growth. Activating mTOR with performance-enhancing drugs could do the opposite, increasing the growth of new tumours. Even if athletes are happy to put up with this long-term risk for the short-term gain of increased muscle strength many of the activating hormones are multifunctional with possibly more immediate detrimental functions. We know insulin has a key role in controlling glucose metabolism, but IGF-1 is also active elsewhere in the body; it regulates growth and development, especially in nerve cells, as well as having the ability to control cellular DNA synthesis. Perhaps dopers are better suited to stick to the myostatin STOP pathway?

Myostatin: the stop button

Myostatin is a molecule that owes its discovery entirely to the wonders of modern molecular biology. When I first studied biochemistry in the 1980s the discovery of a new hormone signalling pathway would involve painstaking purification and testing of molecules isolated from natural sources—usually the glands of dead animals. Myostatin was discovered as a new member of a growth factor family by a technique called degenerative PCR.

PCR (Polymerase Chain Reaction) is a technique that dramatically amplifies the genetic material (DNA) present in a cell. It is used extensively in modern molecular biology. Without it we would have had no Human Genome Project for example. PCR is key to modern forensic science in identifying murder suspects and in forensic archaeology to explore the DNA of long-extinct species such as the woolly mammoth. Degenerative PCR is PCR done badly. Instead of looking to amplify the sequence you know, you try to select for minor modifications of the sequence. In this way you can forensically 'discover' DNA sequences coding for previously undiscovered proteins that are similar, but not identical, to one you started with. Myostatin was discovered by a team working on the Transferring Growth Factor-β (TGF-β) hormone family.

In the days of my youth once a new gene was discovered an attempt would be made to purify the protein product it coded for and study it in the test tube to see what function it might have. However, the modern trick is to go right ahead and remove the gene completely from a mouse; the hope being that these gene knock out mice would lack some function that would make the role of the gene obvious. Now these brute force whole animal experiments frequently don't work as expected. Many's the time a group of researchers sit in front of a brood of perfectly happy mice running around without a single copy of their favoured gene, scratching their heads and wondering what to do next. Nature frequently has a built in redundancy so it can compensate elsewhere for the lack of a gene. Still it is a rather deflating moment when you find a mouse can function happily without any help from the gene you have spent your scientific life studying.

In this case the myostatin team uncovered the best of all possible results.[4] The knock out mice lived but with an altered biology that clearly pointed to the role of myostatin in the body; they had between two and three times as much muscle mass as the control mice, due to both increased numbers and size of muscle cells. It was then shown that breeds of cattle known to be genetically 'double muscled' (the Belgian double muscled cattle, illustrated in Figure 12) were shown to be deficient in myostatin.[5] The final piece in the jigsaw came when a human child with a dramatic increase in muscle mass at age five, was shown to have a defect in his myostatin gene.[6] Myostatin was unequivocally identified as the STOP pathway in muscle growth. Removing it or inhibiting it increases muscle mass even in the absence of excessive weight training.

So what are the details of the myostatin pathway? Do they indicate obvious sites for dopers to develop drugs to increase muscle mass? The

Figure 12 A bull with a genetic defect decreasing myostatin production
Note the overmuscling effects of turning off the STOP pathway

first thing to note is that unlike mTOR, myostatin signalling is specific to the muscle. This is a good start for any prospective dopers as a designer molecule that affects myostatin will be targeted to the right place. Even more usefully the myostatin pathway encompasses a range of possible doping targets other than myostatin itself. To become effective in the muscle cell, myostatin needs to be modified by other proteins. These proteins, called proteases, break myostatin down to make it an optimum size for activity. Once active, myostatin binds to proteins on the muscle cell surface that bind to the TGF-β hormone. Myostatin binding activates these TGF-β receptors allowing them to phosphorylate two proteins in the cell called SMAD2/3 and SMAD4. These then move to the cell nucleus, bind to specific parts of the DNA and, by an, as yet unknown, mechanism, switch off muscle protein synthesis (see Figure 13).

A careful look at Figure 13 reveals that a lot of things have to be right for the muscle to turn off its protein synthesis. Myostatin needs to be expressed in the cell; it needs to be processed by proteases; it then needs to travel in the blood and bind to its receptor; the activated receptor must then phosphorylate SMAD2/3 and SMAD4 which—finally—must go to the nucleus and turn off muscle protein synthesis. To make matters more complex the cell produces a number of inhibitory proteins. Some of these bind to myostatin and stop it working and others, like a protein called SKI, bind to SMAD and do likewise.

Ergogenic aids: biochemical optimisation of weight training

Let's assume—like most sports people—that you are an athlete more concerned with strength rather than looks. You are not after the perfect figure, you just want to win that gold medal. There is no substitute for time spent in the gym. Although some drugs do increase muscle mass at rest, the increase is trivial compared to that achieved by strength training. Don't think for a minute that someone found guilty of anabolic steroid doping hasn't also put the requisite number of hours in at the gym. Intense weight training is the key to increasing muscle mass. Are there biochemical tricks we can use to help us in the gym that don't trigger the ire of the anti-doping authorities?

Myosin and actin are proteins. To increase muscle power therefore requires an increase in protein synthesis. Does this mean we need to eat more protein in our diet? In aerobic sport we discovered how critical it is to control our carbohydrate diet before a big race. Is the protein intake in the diet as important for optimal performance in power events? In one sense this is trivially true. Our metabolic pathways cannot convert our stored fat into protein. So we can't grow our muscles if our diet is protein deficient. But do we need to overcompensate by adding protein supplements to the diet?

The current theory is that new protein is laid down little by little as a direct result of each exercise bout.[7] There is no long-term chronic effect on muscle metabolism. Consequently your fortieth session will have pretty much the same effect as your first. Does this mean that you need to continually take in extra protein to feed this growth? Body builder

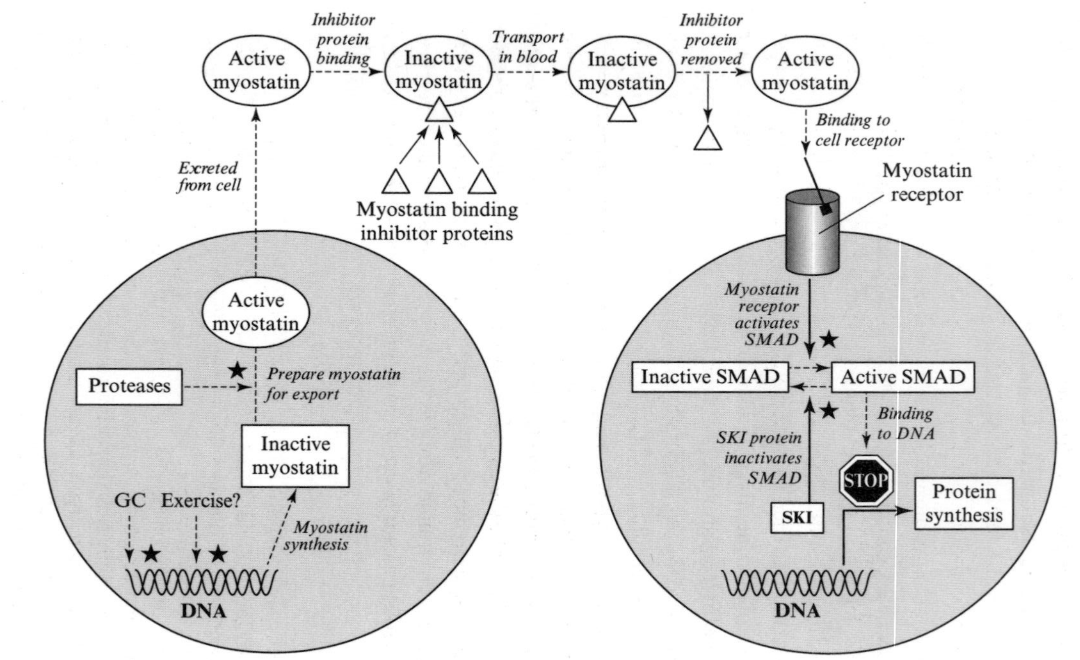

Figure 13 The myostatin 'STOP' pathway for turning off muscle protein synthesis

and weight lifting web pages would have you believe this—they are full of adverts for protein shakes, designed to optimise the laying down of new muscle cell protein. However, much of the increase in protein uptake in muscle is due to a decrease in the breakdown rate, rather than an increase in new protein synthesis. Therefore although it is necessary to modify your diet during weight training—you are aiming to put on weight after all—adding protein drinks alone is not optimal. It is beneficial to combine them with carbohydrates and even fats. In my mother's day (and I guess evolutionary prehistory) this was called a balanced diet. Despite the mass marketing of drinks to enhance muscle strength training, there is no nutritional 'quick fix' in strength-based sports to match the magic bullets of carbohydrate loading or sugar drinks for the long distance runner. To return to our construction analogy increasing the number of protein bricks doesn't increase the speed that the house is built.

What about the type of 'brick'? It is suggested that some sources of protein are better than others. Specifically animal protein sources are claimed to be better than vegetable ones. Curd cheese outperforms bean curd. What might be the mechanism for this difference and can athletes take this to an extreme for added benefit? Proteins are made up of over twenty different amino acids linked together in a specific sequence. When we eat protein in our diet it is broken down in the stomach and

---

A simplified picture of events leading to the turning off of muscle protein synthesis by myostatin. A signalling cell produces myostatin in response to signals such as glucocorticoids (GC) or (possibly) exercise. The inactive protein is then processed (shortened by proteases) for removal from the cell. Outside the cell it circulates in the blood in an inactive state due to binding to inhibitor proteins. Once these are removed active myostatin binds to its target muscle cell and activates muscle protein synthesis by a phosphorylation cascade ending up with a protein called SMAD binding to the DNA. However, other proteins—such as SKI—can inactivate SMAD and so prevent myostatin inhibiting protein synthesis. Even this highly simplified version of this pathway contains a number of unknown molecules and interactions. Altering any one of these has the potential to dramatically alter muscle protein synthesis, leading to the increase in muscle size shown in Figure 12.

intestine and absorbed as the constituent amino acids. These are then transported around the body and, eventually, rebuilt into new sequences that make up the actin and myosin protein filaments necessary to build up muscle mass.

Maybe the differences between milk and soya lie in differences in the nature of the amino acids present. Could some type of amino acids be better able to get into muscle cells? The scientific jury is still out on this; the best theory is that the key is in the absorption rates from the gut into the bloodstream, rather then the rate of absorption into the muscle cells themselves.[8] Anabolism is the building up of molecules and catabolism their breakdown. Milk proteins are more slowly absorbed on average than soy and wheat proteins. Therefore an increase in the diet feeds through smoothly to the body. Cell processes have time to adjust and the increased protein flows smoothly into the muscle cell. However, soya protein is absorbed more quickly. The short, sharp shock of a rapid amino acid rise in the blood from the soya protein breakdown sends the chemical message that the body has too many amino acids. Emergency signalling pathways are initiated and the body responds by sending the resultant amino acids to the liver for disposal (catabolism), rather than the muscle for growth (anabolism). The slower delivery of the less readily absorbed milk proteins therefore results in more amino acids being available for muscle growth.

There are echoes in this argument of the debate about the healthiness of carbohydrate food intake. Carbohydrates that are rapidly absorbed from the gut are quickly converted to high sugar levels in the blood. These are called high glycemic index (GI) foods. Eating high GI foods such as white rice and French baguettes are said to trigger less desirable physiological consequences than eating low glycemic index foods, such as brown rice and whole wheat bread. For both carbohydrates and protein, it seems that slow digestion is the key to the most efficient and healthy use.

If the difference between milk protein and soya protein is in their relative absorption in the gut, then the fine details of the amino acid composition is irrelevant. Once the proteins are broken down into their constituent amino acids, the muscle does not care about the relative ratios of the twenty amino acids. Indeed given the web of metabolic pathways able to interconvert most of the amino acids in the body it seems unlikely that relatively minor changes in amino acid composition of a protein could cause significant differences in muscle protein synthesis. The corollary of this is that there is no magic mixture of amino acids that can rapidly enter the muscle cell and enhance muscle growth. A protein shake just needs to ensure that it contains the proteins in the right form to be absorbed slowly in the gut.

Ergogenic aids: amino acids as a poor man's anabolic steroid?

So if a balanced diet (maybe with a slightly raised protein level) is all you need to optimise your weight training, why are there so many advertisements for wonderful mixtures that will enhance your training performance? The web and health food stores are full of supplements to enhance the effects of weight training. Among the many products touted are ergogenic aids that contain large amounts of individual amino acids. Can we discount these products out of hand given our knowledge about a balanced protein intake? Perhaps not. These products are by and large not trying to subtly alter the body's amino acid dietary composition. Instead they are attempting to hijack cell signalling pathways. This is akin to how anabolic steroids work. Biochemically speaking, people taking these products are trying for the same short cuts, using high doses of 'natural' compounds' that are both cheaper than black market steroids and also less like likely to be banned by sporting bodies. The manufacturers' claims are impressive: different amino acids can stimulate hormone release, activate metabolic pathways and reduce the natural rate of muscle protein breakdown; even the immune system can be improved so that nothing gets in the way of your trip to the gym.

So is there any evidence for these specific amino acid effects? Many of them claim to be able to perturb hormone signalling pathways. Insulin for example, is not only a hormone that controls the level of our blood sugar; it is also able to improve amino acid uptake into muscle cells. Both insulin and human growth hormone have been suggested to be anabolic (muscle building) signals. Weight training programmes therefore promote amino acid supplements as specifically being able to 'naturally' increase the body's concentration of these hormones. This is not without some scientific basis. The amino acids arginine, lysine and ornithine can all increase the release of growth hormone from the pituitary glands. Arginine can also increase the release of insulin. However, the doses given in the supporting scientific experiments are so high that they can only be achieved by intravenous injections of the relevant amino acids.

Take arginine for example. The supporting scientific data suggests 30 g is required. Arginine supplements are usually taken in 1 g pills with a suggested maximum of 2 g/day. Whilst oral doses of up to 10 g are just about tolerable any more causes what scientific papers euphemistically describe as 'gastrointestinal distress'. And 10 g is not enough to have a positive effect on hormone release.[9] While, the amino acid ornithine can release even higher doses of growth hormone than arginine, again this is not possible at the one to two gramme daily doses that can be taken orally.[10]

Insulin levels are well controlled. Even the spike you get with high GI foods comes down to normal levels fairly quickly. The upshot of all this is that it is not possible to dose yourself with enough of a manufacturer's pills to reach the blood amino acid levels that scientific research shows is necessary to induce hormone release. In the case of growth hormone sixty minutes of moderate intensity exercise releases more growth hormone than any amino acid supplement ever could, even if they are taken at very high doses.

What about other amino acid supplements? We have already mentioned the three branched chain amino acids (BCAA) in Chapter 3 as possible modifiers of brain fatigue during long distance aerobic exercise. More recent claims suggest they can also improve the benefits of weight training. But the evidence is weak. It merely involved adding

BCAA to tissue extracts outside the body and showing an increase in protein synthesis and a decrease in protein breakdown. No one has been able to reproduce anything approximating this effect in a real living person.[11]

However, before we leave the realm of amino acid supplementation one product deserves mention. It is not an amino acid, but a breakdown product of one. Beta-hydroxy beta methylbutyrate (HMB for short) is synthesised in the body from the amino acid leucine. It is unclear whether it plays a normal role in cell function. However, what does seem clear is that supplementation at the 3 g per day level achievable by oral dosing does have a small effect on muscle strength. The mechanisms underpinning these effects are unclear at present.[12] We don't know whether this will prove to be a small obscure research finding that leads nowhere in particular, or whether it signals the dawn of a new class of compounds that can enhance the efficacy of strength training.

Ergogenic aids: creatine the wonder pill?

So much for amino acid supplementation. We now turn to a molecule that we first met when we discussed energy metabolism and aerobic running. Creatine was shown to have a positive effect via increasing the efficient use of energy to drive short-term repetitive sprints. However, it also appears that it is able to improve the power of those sprints by enhancing muscle hypertrophy. A strength-training programme seems to work better when combined with creatine. Again as for sprinting the doses used for strength training, though safe, are far above what we can get in our diet. The mechanism is unknown. Still in the words of one scientific review: 'In summary, the predominance of research indicates that creatine supplementation represents a safe, effective, and legal method to enhance muscle size and strength responses to resistance training'.[13] In researching this chapter it was a pleasure to finally come across a sentence like this, where good science underpins the advertising hype. It is a rare phenomenon.

How is muscle strength increased with ergogenic aids?

There are therefore only three currently well-validated methods for increasing muscle strength that do not arouse the wrath of the doping agency: resistance (weight) training coupled with a balanced increase in calorific intake; creatine supplementation; the intake of beta-hydroxy beta methylbutyrate (HMB). Does our trawl through the mTOR and myostatin pathways indicate the mechanism for how these methods might work? With regards to strength training, the induced muscle hypertrophy is a very local event. This is likely to be even more so for the extreme muscling seen in elite athletes and bodybuilders. It is not likely that general (system-wide) signalling events are as important as events at the muscle itself.

So how does exercise affect the GO (mTOR) pathway as it works in muscle? Let's revisit this pathway (see Figure 11). There seems to be a direct mechanical effect on activation of mTOR that does not require insulin or insulin-like growth factors at all. Being mechanical the effect could naturally be restricted to the mTOR present in muscle cells. But exercise also increases the production of growth hormone in the brain. This goes on to increase the production of IGF-1 and insulin throughout the body. Both insulin and IGF-1 can activate the m-TOR pathway. How could this hormonal activation of mTOR be targeted to the muscle alone? Here the somewhat elusive muscle-specific special growth factor MGF is likely to be key.

What about the STOP (myostatin) signal? Again there is some uncertainty. It is clear that inhibiting myostatin enhances muscle mass. Some studies show a drop in myostatin after weight training exercise. But this effect is not universal. The problem may relate to the difficulty in studying local signalling pathways in humans. Local pathways require local samples. For muscles this requires a biopsy of the muscle tissue. Many human studies rather naturally restrict the number of biopsies. However, any increase in myostatin must be time limited. What goes up, must come down. Therefore as we don't know the speed of the rise and fall, a single timed sample may be taken too early or too late to measure an observed effect.

The solution is to take multiple time points. In 2007 in a heroic study (especially for the volunteers), Scott Trappe at Ball State University in Indiana, USA studied the response of muscle genes to exercise.[14] Eight muscle biopsies were taken: immediately pre-exercise, immediately post-exercise and then 1, 2, 4, 8, 12 and 24 hours later. A clear myostatin fall was noted, the drop being largest at 8 hours. Such a fall could be the trigger for increased muscle protein synthesis.

Intriguingly other studies have shown that during the rest between exercise bouts myostatin levels actually *increase*.[15] This confirms the earlier work described in this chapter. There is no long-term chronic effect on resting muscle metabolism, what matters happens immediately post-exercise. New protein is synthesised as a direct result of each exercise bout.

So in summary the best current data suggests that weight training both increases the ON pathway and decreases the OFF pathway for muscle cell growth. What about creatine and HMB? Creatine is non-toxic and cheap and consequently easy to use in scientific studies. Therefore it is somewhat surprising that the mechanism of creatine action is almost completely unclear. Although creatine supplementation has been shown to increase acute weightlifting performance and training volume, the positive effect goes beyond this.

Creatine could theoretically work by enhancing cellular energy reserves or modifying signalling pathways. Although creatine is a molecule that is involved in energy metabolism, it would be surprising if a lack of ATP in the muscle cell restricted the ability to grow, especially as the growth occurs *after* the exercise has stopped. On the other hand a study in 2010 by the Iranian scientist Saremi intriguingly suggested that creatine is able to decrease myostatin levels in the blood[16] suggesting the possibility of creatine effects on signalling pathways involved in protein degradation.

Speculation about the role of HMB is more fruitful. It can cause an increase in the synthesis of cholesterol, which is needed for muscle repair after intense training; it can also affect cell signalling either via activating the mTOR GO signal as well as interacting with an alternative to the myostatin STOP pathway—the ubiquitin pathway. The mTOR mechanism is perhaps the most compelling. HMB increases muscle protein synthesis in mice. This increase is prevented if a molecule is added that inhibits the mTOR pathway namely rapamycin.[12] This simple experiment suggests that HMB works via its effects on the mTOR pathway.

How is muscle strength increased with doping?

In the next chapter we will explore this question in more detail. But it is worth making a few pertinent points here while the mechanisms for muscle growth are fresh in our memory. First of all it is clear that science is learning a lot more about the key signalling pathways. As well as mTOR and myostatin, there are others that I have not had time to describe in detail such as the ubiquitin system, the FOX/forkhead pathways, NF-kB, integrin and calcineurin. These all provide possible targets for dopers. As we learnt in earlier chapters the amount of resource society puts into scientific research in these areas largely depends on how important they are in human disease. Muscle damage and wasting are key to a range of diseases such as muscular dystrophy, motor neuron disease, peripheral nerve damage and of course the ageing process itself. So there is likely to be a commercial drive to make new pharmaceutical drugs.

When it comes to pathways to target most drugs are enzyme inhibitors. It is much easier to stop a pathway then start one. Turning on a pathway requires many different enzyme systems to be activated. Otherwise you just convert a bottleneck in one part of the pathway to a bottleneck in another. For example when we start running, as many as ten enzymes are activated to cause the almost twenty-fold increase in muscle oxygen consumption observed. In contrast just one inhibitor can stop oxygen consumption dead—that is why hydrogen cyanide is toxic for example. As it is with oxygen consumption so it is likely to be with muscle growth. Any block on the myostatin pathway will prevent protein breakdown; the added advantage is that the pathway is muscle-specific. There is a wide range of proteins that have to be in the right place at the right time to keep myostatin active. Just one break in this complex web of interactions has the potential to lead to an increase in muscle cell growth.

The next generation of dopers are likely to target myostatin.[17] Indeed you can't access a bodybuilder internet chat room at the moment without someone asking where they can get hold of a myostatin inhibitor. But before we delve into these new 'smart' drugs, we should visit the old ways first. Any discussion of doping methods is going to have to deal with anabolic steroids in detail sooner or later. That time has now come.

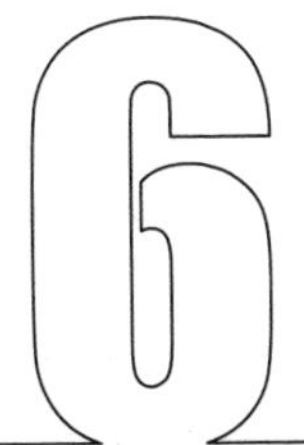

Steroids And Beyond...

'Steroids can help a female sprinter to lower her 100 m time by about four tenths of a second or four metres faster. The effects of steroids upon male 100 m sprinters are about two tenths of a second or two metres faster.'
Victor Conte

What is a steroid?

We now turn to the most infamous doping agents of all: anabolic steroids. A steroid is defined by its chemical structure: in this case four rings (see Figure 14). The chemical nature of these rings, and what is stuck on to them, then determines its biological reactivity. Small chemical differences can have dramatic effects. The fine details of the structure matter; moving just a single oxygen and hydrogen atom changes the male sex hormone testosterone into the female hormone estradiol, converting a hormonal message for higher muscle mass, deeper voice and thicker hair growth into one that reduces muscle mass and encourages breast development.

The similarity in chemical structure is beneficial to the body. Rather than having to evolve completely different pathways for making male and female sex hormones, minor adaptations of existing pathways are all that is required. This is possible because signalling works on the 'lock and key' principle described in the previous chapter; the body is happy to use the same material to make a whole range of keys. Why make house keys out of different materials when a small change in shape is all that is needed to make it specific for an individual house? Similarly in biochemistry, the same basic structural intermediates can happily co-exist in the cell, as only those with the precise shape can open the lock and elicit the correct response. Steroids are therefore used for many cellular processes other than sexual signalling.

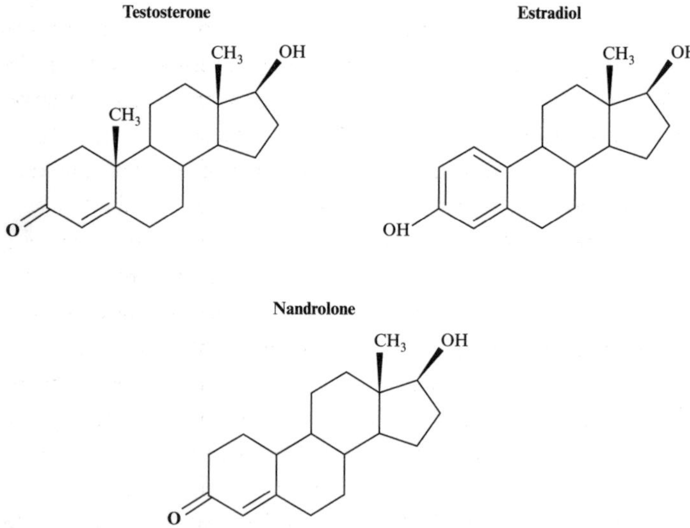

Figure 14 Spot the difference

Difference between the structure of the male sex hormone (testosterone) and the female sex hormone (estradiol). A single enzyme, aromatase effects this conversion. The synthetic anabolic steroid, nandrolone, is even more closely linked chemically to estradiol. The key difference is the addition of a hydrogen (H) to the oxygen (O) on the bottom left of the structure.

The first molecule synthesised in the body that contains the four-ring steroid structure is lanosterol. All steroids are therefore derived from this molecule. While it is easy to draw structures on paper it is not so easy in real life to make the requisite chemical changes. It takes nineteen chemical reactions to convert lanosterol to a steroid molecule we are all familiar with—cholesterol. Doing this synthesis in a laboratory would take many weeks of hard graft and probably require high temperatures, or extremes of acid or alkali. The body's enzymes manage it at room temperature and neutral pH in seconds.

Cholesterol is known to most of us for problems associated with abnormal deposition of its metabolites on artery walls, the resulting 'furring up' shrinks the vessels and increases the risk of heart attacks and strokes. However, cholesterol also plays a key role in maintaining the fluidity of our cell membranes—brain signalling for example requires a cholesterol-rich insulating sheath to allow for more efficient transport of the electrical signals. When this cholesterol sheath breaks down, as in diseases such as multiple sclerosis, brain signals become erratic with debilitating effects on the body's function.

Cholesterol can be converted into a number of useful compounds. The liver turns cholesterol into bile salts, which are stored in the gall-bladder. These salts are then released into the intestine when we eat a meal, where they aid the digestion of fatty foods. This role does not require chemical alteration of a molecule—it is purely a physical effect. Steroid molecules are hydrophobic; they hate water and do not mix with it. But they can mix well with other oily, fatty molecules. Cholesterols are therefore able to dissolve (or more technically emulsify) the fat in our diets and enable its efficient absorption by the body. This is why, if you have your gall bladder removed, you have to be careful how much fat is in your diet. The fat you eat will not mix with the water in your gut and, in the absence of bile salts, will be poorly absorbed.

As well as playing these key structural roles, cholesterols can also play a role in signalling. Cholesterol molecules bind and modify a class of proteins involved in limb development. These proteins were discovered in the fruit fly and the class of proteins was called 'hedgehog', because the loss of the gene resulted in the fruit fly embryos being covered with small, spiny projections. In humans there are three versions of the *hedgehog* gene, named after three types of hedgehog: *desert* and *Indian* named after animals and *sonic* named after a video game character. Cholesterol plays the role of binding to the *sonic hedgehog* protein and transporting it around the body. However, this is a minor part of cholesterol's signalling role. More important is that cholesterol is the immediate precursor of all of the steroid hormones (see Figure 15). These hormones have a variety of functions in the body.

Glucocorticoids are involved in stress responses and control blood pressure and the immune system. The most famous example is the naturally occurring cortisol. Cortisol is available as a cream to treat

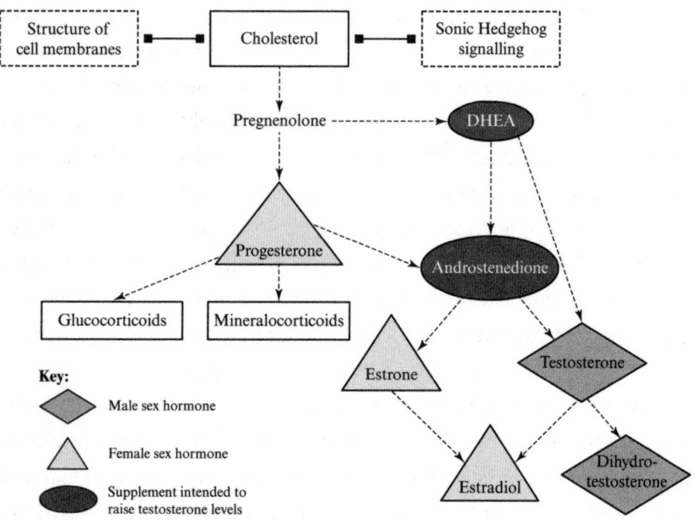

Figure 15 Making a sex hormone from cholesterol

The arrows indicate the metabolic pathways—simplified in this diagram. Note how closely linked are the manufacture of the male and female sex hormones. Indeed both males and females have levels of progesterone and testosterone respectively, though at much lower levels than those of the 'correct' sex. It is problematic for athletes to affect male sex hormone levels without having some effects on female hormones.

skin allergies; synthetic versions of the molecule (e.g. prednisone) are used to treat asthma. Corticosteroids are banned for Olympic sport when injected, taken orally or inhaled, except for medical reasons such as for asthmatics. However, this ban does not relate to any purported effects on muscle strength. Indeed as we saw in the last chapter glucocorticoids actually increase production of myostatin, so are likely to turn off muscle protein synthesis. However, as any severe asthmatic will tell you, having to take high dose corticosteroids orally has profound mood altering effects. It is this possible psychological benefit that the anti-doping authorities aim to curtail.

Mineralocorticoids, such as aldosterone, controls salt metabolism and blood pressure. They have never been banned, though interestingly this blood pressure control system is linked biochemically to one of the genes—called *ACE*—that can alter human exercise performance.

We are then left with the three classes of sex hormones, two female (**progestagens** and **estrogens**) and one male (**androgens**). Progestagens, such as progesterone are largely responsible for maintaining pregnancy. Estrogens, such as estradiol, are involved in the development of what we would think of as female secondary sexual characteristics, breast development for example. Interestingly although men have somewhat lower levels of estrogens than women, they still need some of this so-called 'female' sexual hormone for normal sexual function. But as no one has found a sport where female hormones can enhance performance, there is no fear of their use as doping agents.

This of course is not the case for androgenic (male) hormones. To be even more specific what interests the doping agencies are **anabolic androgenic steroids** (hence the abbreviation AAS). These are steroid hormones that control both male sexual function (androgenic) *and* have the ability to build up big molecules from small ones (anabolic). The anabolic process is particularly targeted at increasing muscle protein resulting in larger, more massive muscles. Anabolic steroids are therefore banned in all forms, natural or synthetic.

There are five frequently discussed naturally occurring androgenic steroids. During development **testosterone** is responsible for the creation of male primary sexual characteristics (the reproductive organs) while **dihydrotestosterone** is more closely linked to secondary male characteristics (facial hair etc.). These compounds are both anabolic i.e. they can build up muscle mass. A third testosterone derivative—**epitestosterone**—is not anabolic; indeed it can behave as an inhibitor of testosterone's activity. This leaves two molecules **androstenedione** (known colloquially as 'andro') and **dehydroepiandrosterone** (DHEA). In contrast to testosterone, which is produced predominantly in the testes, androstenedione and DHEA are mostly produced in the adrenal glands. While clearly androgenic 'male' hormones it is not clear that they are anabolic (muscle building). DHEA and 'andro' have been the source of much controversy over the years; they are currently on the banned list.

Cellular mechanism of testosterone action

How does testosterone work to increase strength in the body and does taking more make us stronger? The salient details are illustrated in Figure 16. Steroid hormones generally make new signalling proteins, rather than modifying the function of existing ones by phosphorylation. How do they do this? As we have already noted all proteins are made from reading out the genetic code on our DNA. Steroid hormones determine which code should be read, and therefore which protein is synthesised. They do this by binding directly to the DNA in the nucleus. Or rather, because nothing in biochemistry is that simple, they bind to molecules which then bind to DNA.

The intermediary molecules in question are testosterone receptors and they exist in all our cells. They become active only after binding to testosterone. This binding sends the receptor to the cell nucleus. Inside the nucleus is the DNA, which codes for the genes that control the production of the body's ≈20,000 proteins. Only specific bits of DNA can bind to the activated testosterone receptor. These specific DNA sites are called hormone response elements (HRE).

Once the testosterone receptor binds to the HRE, a gene is activated. This gene, in turn, makes a protein which can then change the activity of the cell. Because there can be many different hormone response elements, all responding to different extents, one hormone can have a multitude of different effects in a cell. Indeed the effect can be different in different cells depending on whether the DNA containing the HRE is made accessible to the hormone, or blocked, perhaps by the effects of separate hormones. It seems biochemistry weaves a very tangled web at times; but of course there is a method to this complexity. You don't want testosterone activating protein synthesis in just any old cell. So you prevent this by a variety of means: blocking its uptake in the cell;

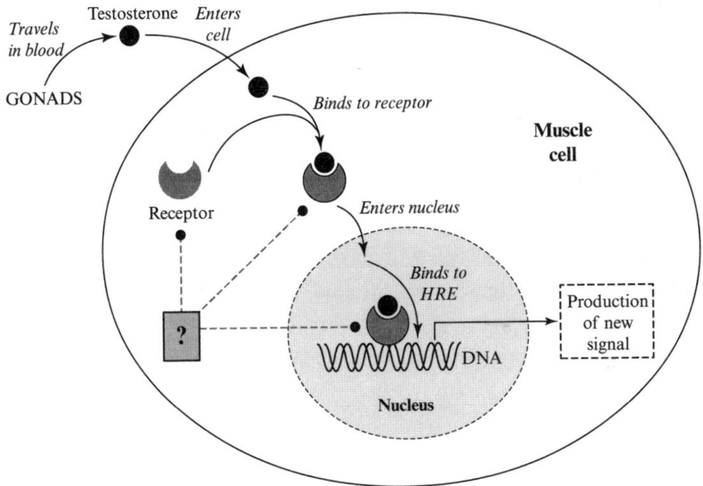

Figure 16 How do steroid hormones work?

Testosterone is produced in the gonads (testes) and travels around the body to its target cell—in this case a muscle cell. Upon entering the cell, it encounters a receptor. The testosterone/receptor complex enters the nucleus and binds to a hormone responsive element (HRE) where it can affect the synthesis of products that can control cell function. The exact details of the hormone responsive elements are unknown, as are the nature of any proteins (indicated with a ? on the diagram) that can destroy the receptor or affect its binding to testosterone. These are all potentially uncharted areas for doping to enhance testosterone efficacy.

not synthesising its receptor; modifying its hormone response element. Of course the opposite effect will accentuate testosterone's protein synthesis effects—exactly what the dopers would want. As scientists understand more about the mechanism of these hormone responsive elements a multitude of pharmacological options are likely to open up for sports doping.

So what kinds of proteins are controlled by the HREs for testosterone and how do they interact with the pathways for muscle growth that were described in the last chapter? In fact it is surprisingly difficult to get a good answer to this. The scientific literature is skewed by medical research and the resulting associated funding. Most funded research focuses on detecting the targets of testosterone in the prostate, rather than the muscle. This is because the growth of prostate cancer is accelerated by testosterone. Indeed an initial screening for prostate cancer is to measure a protein that is activated by testosterone binding to an HRE in the prostate. This protein is called PSA—prostate specific antigen. PSA screening is routinely carried out on all middle-aged men in the USA. Measuring its levels in the blood gives a good indication of abnormal testosterone activity in the prostate.

Presumably testosterone will activate different HREs in the muscle cell than it does in the prostate. After all male prostate glands do many things, but increasing in size after exercising is not one of them. An activated testosterone HRE in the muscle cell should result in the production of a new protein that can signal growth of the cell. It is not too difficult to get a list of possible candidate growth factors that supplement manufacturers think could be responding to testosterone in the muscle. IGF we have met before—insulin or mechano growth factor regulates the mTOR pathway. FGF is fibroblast growth factor and might play an accessory role in the production of 'satellite cells' that feed muscle cell growth. But other growth factors such as a nerve growth factor (NGF), epithelial growth factor (EGF) and connective tissue growth factor (CTGF) are not likely to enhance muscle growth.

Growth factors are big business outside sports doping as well. They are used in cosmetic products to allegedly enhance the recovery of skin cells following the trials and tribulations of being exposed to the elements. At least with skin products there is a chance—albeit small—that these growth factors could access the cells they are intended to affect.

There is very little hard evidence that any of the growth factors mentioned above are activated by testosterone. The one small piece of data there is hints at the presence of a testosterone-binding site on DNA that activates IGF-1 expression. Although this has not yet been shown to be active in muscle cells, the discovery is intriguing. In the last chapter we saw that IGF-1 activates the mTOR pathway. This would suggest that testosterone activates the GO (mTOR) pathway of muscle cell growth, rather than inhibiting the STOP (myostatin) pathway. Is there any evidence for this from studies in real people, rather than in test tubes?

Does testosterone work?

Attempts at increasing testosterone levels have a long history of variable success. In males testosterone, as its name suggests, is mostly produced in the testes. It has long been known that removing the testes—castration—has dramatic effects on male attributes. Records stretch as far back as the Assyrians, but given its importance in food production and animal behaviour, castration almost certainly developed hand in hand with the domestication of animals. Aristotle (300 BC) speaks of castration of humans (making them eunuchs) and chickens (turning cocks into capons). In the 1920s the famous US mail order firm Sears, Roebuck & Co. even sold home 'Caponizing Sets' with full graphic instructions; you will be pleased to note that readers were advised to practice on dead birds first.

Capon meat is considered juicier and more tender than cock's meat. Some even prefer it to the female chicken. What is the reason for this difference? In true Aristotlean fashion, a declaration was made unsupported by evidence. 'Q: Why is a capon better to eat than a cock? A: Because a capon loses not his moisture by treading of the hens.'[2] The modern view is more prosaic. Capons taste different than cocks even if neither has seen a hen. In the case of taste exercise is bad for you. Castration causes behavioural changes that reduce the amount of exercise; this increases the fat content of the capon and makes for a juicier bird.

While the meat industry sought to make docile fat animals, not surprisingly the health and sports industry sought to do the reverse. If removing testicles made animals docile could adding them make them more aggressive? Could this extend to humans? In the absence of knowledge about surgically adding functional testicles, people at first tried more simple methods. Eating testicles is still common in Oriental cuisine and modern day Testicle Eating festivals have sprung up all over the USA. Whilst the testosterone 'kick' you can get this way is miniscule, the science behind the idea is robust. The testicle does indeed produce a compound that affects male function.[3] The pioneering work in this area was done by the eighteenth-century Scottish surgeon John Hunter, but the first published experiments were undertaken at Göttingen by the German scientist, Arnold Berthold. In 1849 Berthold successfully transplanted testes into the abdomens of capons. Amazingly this reversed many of the effects of castration. Scientists were astounded that the effects persisted despite there being no physical connection to the nervous system; the transplanted organ must have made a magic molecule that travelled around the body. Thus the science of hormones (endocrinology) was born.

By the end of the nineteenth century scientists judged, correctly, that whole testicles would contain only a small amount of the active compound, which might anyway be destroyed by the process of digestion. Therefore they started to make a serious attempt to introduce the 'male chemical' directly into the body.[3] In 1889 a French doctor, Charles-Édouard Brown-Séquard injected extracts of dog and guinea pig testicles into other animals (including himself). All sort of claims were made for the success of these treatments in rejuvenating health. Some even claimed to have found the secret of eternal youth. As Groucho Marx sings in *The Cocoanuts* 'If you're too old for dancing/Get yourself a monkey gland'.[4] A small cottage industry was built up injecting testicle extracts into people. In 1920 A Russian-born French surgeon, Serge

Voronoff, went one stage further and mimicked Berthold's experiments by grafting thin slices of chimpanzee testicles into people's scrotums. Sport was not immune. Remember the 'monkey gland' 1939 FA Cup Final between Wolverhampton Wanderers and Portsmouth discussed in Chapter 1? The chemist responsible, Menzies Sharp, was reputedly a disciple of Voronoff.

The idea of gland and gland extract injections came into disrepute, as the results did not appear to match the hype. A more potent product was needed. The hunt for the active compound needed large quantities of starting materials.[3] Fred Koch produced a highly active preparation from a ready source—the infamous Chicago stockyards which produced over 75 per cent of the meat eaten in the USA at the time. Twenty kilogrammes of bulls' testicles were converted into 20 mg of active hormone extract; a one million-fold purification. Not to be outdone, in Europe, Adolf Butenandt was busy purifying human hormones from the urine of German policemen, 15,000 litres of it to be precise. Fifteen milligrammes were recovered. Obtaining enough pure compound to be injected into people was proving a challenge on an industrial scale. But there was big money to be made. In the 1930s three European pharmaceutical giants threw their hats into the ring, the Swiss (Ciba), the Germans (Schering) and the Dutch (Organon). In the following twenty years the field of steroid biochemistry grew apace. The natural steroids testosterone, androstenedione ('andro') and dehydroepiandrosterone (DHEA) were discovered. But once they discovered how easy it was to artificially modify the basic steroid backbone, organic chemists did not stop at these natural products. New molecules meant new patents meant more money. The range of steroid products athletes use for doping is a direct result of these pioneering European pharmaceutical companies. The sporting 'legacy' is but a small pimple on the back of this corporate medical research.

The granddaddy of all anabolic steroids is testosterone itself. So does it really work as a purified extract? The key experiments were done in the late 1930s when injections of purified testosterone were shown to have anabolic effects in castrated dogs. Some men are unfortunately born eunuchs due to a genetic abnormality; similar results were found when testosterone was tested on this group of people. In terms of human performance the first study was in 1944 when six older Americans were treated with testosterone.[5] Both the extent and duration of strength-based tests increased compared to placebo. But this was a group who were older (aged 50+) and had already complained of fatigue. Clearly these results spurred on the hormone therapy market, but what about sports? Does testosterone work on fit young males? Certainly word got around that it might. The history here is cloudy. Although early tales of drugged up Nazi soldiers have been discredited, no one quite knows who were the first users of steroids. However, whether it started in the Californian bodybuilders of the 1940s or Soviet weightlifters in the 1950s, it was clear that by the 1960s anabolic steroids were widely used in strength-based sports in many countries.

It was generally assumed that because studies clearly showed that testosterone worked in hormonally deficient or aged populations, they would be of benefit to all users. We have noted before how difficult it is to do controlled experiments with performance enhancing drugs. Still it is surprising to read statements as late as 1991 in the scientific literature such as 'Although anecdotal and theoretical information suggests that anabolic androgenic steroids have positive ergogenic properties, the experimental evidence is equivocal'.[6] Remember this is being written in a peer-reviewed scientific article three years after Ben Johnson's drug-tainted 100 m Olympic sprint in 1988. Was Ben Johnson really wasting his time?

Perhaps appropriately given that some of the earliest steroid abuse by athletes originated in the body builders of LA's Venice Beach, it took a careful study from a Los Angeles scientist to finally resolve this issue. Dr Shalender Bhasin investigated 43 young males to confirm what many had expected all along.[7] High dose testosterone injections increase fat-free mass, muscle size and strength in normal men. The mechanism remains elusive though. Users swear that steroids enable them to train longer and harder. However, Bhasin showed that testosterone has effects even when you were not training. In his study strength was increased most by the combination of exercise and testosterone. But taken on its own testosterone was pretty much as effective as exercise. Clearly the ergogenic effects of steroids exist even if you do no weight training at all.

How does testosterone work?

So testosterone supplementation works. In a normal population supraphysiological doses of testosterone do indeed improve the development of muscle mass. What is happening at the biochemical level? Interestingly different effects seem to occur dependent on age. In younger people testosterone increases new protein synthesis, but has little effect on breakdown.[8] In older people the opposite effect is seen. Breakdown is reduced, but synthesis unaffected.[9] Same result—increased muscle protein synthesis—but achieved by a different method. This shows the difficulties in extrapolating clinical studies that focus on steroid hormone replacement therapy and are likely to be well funded, with the smaller number of volunteer studies on fit young males willing to take testosterone for a scientific study. And of course we know nothing about whether elite male athletes behave differently to the average fit young male.

Despite these difficulties a story is starting to emerge, though there are still gaps in the fine details. We learnt in the last chapter that weight training works by decreasing the breakdown of muscle proteins. In contrast anabolic steroids seem to work by increasing the synthesis of these proteins. This is presumably why steroids work even if you are a couch potato, and why they work even better if you exercise. There is a synergy between increasing protein synthesis (the steroid effect) and decreasing protein breakdown (the training effect). Both increase the levels of protein available for muscle fibre growth.

Details are gradually emerging of what is happening at the molecular level. We have already noted that testosterone activates the hormone responsive element, IGF-1.[1] This fits in nicely with the idea we mentioned earlier—testosterone activates the GO pathway for muscle protein synthesis. However, there is probably something else going on. It is likely testosterone does more than just increases the production of IGF-1 (see Figure 17). Testosterone allows a more efficient use of amino acids inside the muscle cell. Rather than leaving the muscle cell, amino acids are kept available; they sit in the cell primed to be incorporated into the synthesis of the new proteins required for muscle growth.[8] The mechanism behind this effect is unknown.

It is important to remain cautious when we try to translate testosterone supplementation to sports performance. We still do not know if

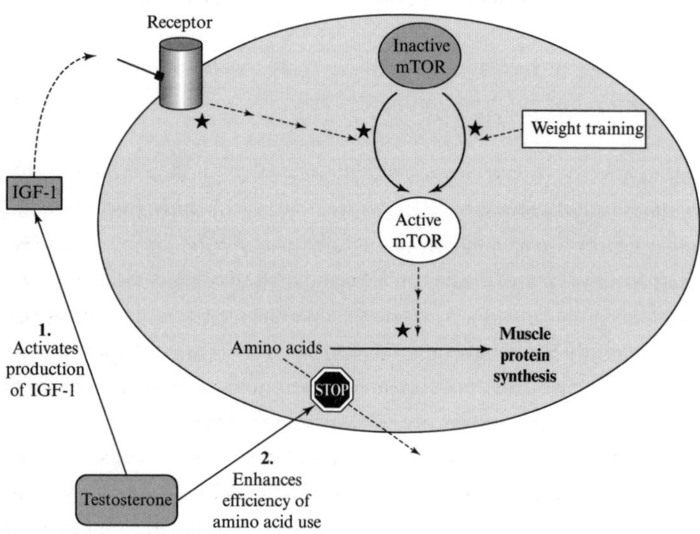

Figure 17 How does testosterone increase muscle growth?

The sum of our current knowledge suggests that testosterone activates the GO pathway for muscle protein synthesis. It does this by activating a hormone response element to increase IGF-1, which then goes on to activate mTOR. However, it also acts by an unknown mechanism to prevent amino acids leaving the muscle cell. This enhances the efficiency of amino acid conversion to proteins. This diagram illustrates two important points. First it confirms the idea that testosterone acts independently of weight training. Secondly, and more importantly, testosterone can act via a number of separate pathways, each of which has the potential to enhance muscle cell growth. This is a potential fertile ground for dopers—or would be if we had any idea of the molecular details of the pathways involved.

steroid supplementation just makes it easier to reach optimum performance or if it actually moves the goalposts to a place they cannot be reached in the absence of steroid enhancement. Put bluntly is it really true that any world record achieved by an athlete in a strength event would be enhanced if that athlete trained on steroids? It seems to me that this is still a live question. There could be a peak of optimum muscle growth. Steroids might make it easier to get there, but the same summit could be reached by anyone if they tried hard enough. Alternatively some athletes could be born with genes that mean that they can reach their optimum strength without the need for added testosterone.

Steroids and the female athlete

Although it is hard to say unequivocally that the fastest man on earth would run faster on steroids, it is much easier to make this statement about the fastest woman. It may shock the scientific purists, but some facts in science (and life) are so clear that they don't need statistics. We don't need the help of a mathematician to tell us that male athletes run faster and throw further than female ones. So giving a female athlete more male characteristics should clearly benefit performance. If this is done early enough in life the effects could be dramatic.

Steroids can change the natural path of development if given before or during puberty. Consequently in the mature adult female there is not necessarily any need to continue the treatment to see a positive effect; a series of judicious childhood injections could last a lifetime. We are talking here about putting the androgen (sex) into androgenic, anabolic steroids. For those of us who grew up watching sport in the 1970s the story is all too familiar. The exemplar is of course the East German state-sponsored drug programme. Long rumoured, its activities were revealed following the fall of the Berlin Wall in 1989. The systematic nature of the doping was revealed by Brigitte Berendonk, a former East German who fled to the West. There she competed in the Olympics as a discus thrower. However, even more significant than her Olympic career was the detailed book she wrote in 1991 entitled *Doping Dokumente. Von der Forschung zum Betrug* (Doping documents—From Research to Fraud). This book was followed up in 1997, by a paper (chapter 1 ref. 8) written with her husband Werner Franke, a Professor of Cell Biology entitled 'Hormonal doping and androgenization of athletes: a secret program of the German Democratic Republic government'.

These works makes sober reading. When the wall collapsed, over 150 classified reports were saved from the shredders, or in some cases sold to newspapers by the dopers themselves. One of these was Mannfred Höppner the top sports doctor in East Germany. In a Berlin court in 2000, Höppner was convicted of being an accessory to the intentional bodily harm of 142 female athletes, including minors. Höppner's Stasi code name—Technik—chillingly resonates with the view of the athletes as a technology to exploit. In a 1977 report to the Stasi, 'Technik' reports that 'At present anabolic steroids are applied in all Olympic sporting events, with the exception of sailing and gymnastics (female), . . . and by all national teams'. The report goes on to add that 'From our experiences made so far it can be concluded that women have the greatest advantage from treatments with anabolic hormones with respect to their performance in sports. . . . Especially high is the performance-supporting effect following the first administration of anabolic hormones, especially with junior athletes' (chapter 1 ref. 8).

This differential effect between male and female anabolic steroid use can easily be seen when looking at an archetypal strength event—the shot put. Over the period 1970–1988, when steroid abuse was widespread in both sexes, there is a far more marked increase in the female best performances (see Figure 18). Out of competition drug testing was introduced in 1988 to reduce the ability of athletes to train at leisure on steroids during their pre-season, coming off them just before an event to avoid testing positive. This female increase drops off when out of competition drug testing is introduced, but there is a minimal effect on male performance.

Apart from the out of competition testing there is another factor that has decreased the use of anabolic steroids by female athletes. Although

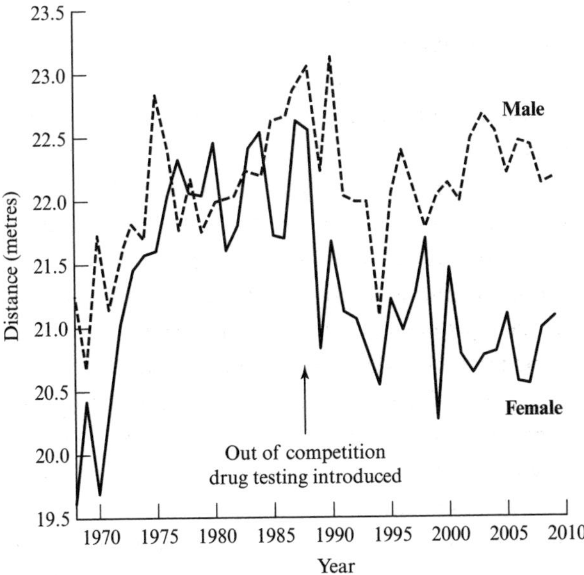

Figure 18 World's best shot put performances for males and females

1968 saw the start of the East German anabolic steroid programme for female athletes. 1989 saw the introduction of (limited) out of competition drug testing to make it harder to use steroids prior to championships. The female curve shows an inverted U shape, suggesting that there was a significant performance effect between 1975 and 1988; all the world's best throws in this time were from Eastern European countries. Point to note: the male shot is heavier which is why the distances thrown are compatible between the sexes.

it would seem to be theoretically possible, no one has been able to separate the androgenic (sex determining) from the anabolic (muscle building) roles of steroid hormones. Increased muscle mass is therefore not the only effects of these hormones. Long-term side-effects include increased risk of cardiovascular disease and cancer, especially of the liver and kidney. In the short term the effects are more sex-specific. In males artificial steroids can be converted to the female hormone estradiol. Therefore paradoxically taking a male sex hormone can lead to the formation of female secondary sex characteristics. The most common feature is gynaecomastia, or enlargement of the breast tissue. Not surprisingly trying to deal with this side effect is a major talking point in the steroid community, especially among bodybuilders. Tamoxifen—a breast cancer drug that reduces female hormone activity—is frequently used by bodybuilders, though the success rates appear to be variable. This illustrates that drug use in sport follows a similar pattern many of us get used to as we get older. We take as many pills to counteract the side-effects of drugs as we do to counter the disease itself.

The female side effects of male sex steroid use are more extensive though, including deepening voice, menstrual problems, clitoral enlargement, extensive growth of facial and pubic hair and male pattern baldness. It is possible that this, as much as the out of competition testing, has resulted in a decrease of anabolic steroid use in female athletes; or at least an attempt to try compounds that can achieve an anabolic effect without the androgenic side effects.

Which steroid is best?

Testosterone itself, although it can be taken in a pill, is quickly metabolised by the liver and inactivated. It can be modified to be less metabolised and more long-lasting, but then needs to be injected. So you then need another set of modifications to make it orally active. Not surprisingly the chemistry of modification is well known, given that testosterone development evolved from a race by organic chemistry companies to make a useful pharmaceutical drug, Over time the pharmaceutical companies have produced a range of products that can be short or long lasting, given as a cream, pill or injection. Dopers take advantage of this to use a variety of steroids depending on the desired mix of these properties. As we have noted before most compounds have been made by companies interested in healthcare benefits. As well as the obvious use to treat patients with low normal levels of testosterone or to help with gender reassignment, anabolic steroids have been used to stimulate red blood cell formation and reverse childhood growth retardation; they have even been experimented with as treatments for chronic wasting conditions such as cancer.

In many cases the clinical use of anabolic steroids has been supplanted by other hormones; these include EPO for red blood cell growth and HGH for childhood growth, both hormone treatments also banned for athletes. The upshot of the replacement of steroids as medical treatments by these newer alternatives is that no major pharmaceutical company has a current steroid development programme. Of course we can't be sure that there are no government based programmes—the East German files contained substances that even in these classified documents were named in code. But it is likely that by and large athletes and coaches are on their own if they want to make new steroids.

Athletes have an additional reason for using novel molecules that the drug companies don't. They don't necessarily need a more effective molecule—they just need one that is equally effective but undetectable in a doping test. This was the rationale for the synthetic steroid THG—or 'The Clear'. The inventor of THG, Patrick Arnold, was a largely self-taught chemist, working with relatively cheap resources compared to a multinational pharmaceutical company. Although undoubtedly talented he is unlikely to be a one-off genius. There is enough information available for any competent chemist to envisage novel steroid modifications. A simple search of the standard scientific literature or a hunt for outdated patents for molecules that didn't quite come up to scratch will usually suffice. Once you have your molecule of choice, basic chemical expertise will probably be enough to synthesis it in the laboratory. Making the process efficient enough to create a pill or injectable product at a high enough concentration for use by an athlete is a bit more challenging. However, designer steroids avoid the real costs a drug company has to incur in product research, animal and (especially) clinical testing. Designer modified anabolic steroids are therefore one of the few areas where I think dopers can effectively create products without piggybacking off new ideas from the medical and scientific establishment. I would be surprised if more products do not appear on the market soon. Perhaps they are already here?

There is not much point dwelling on the fine differences between the various anabolic steroids. At their heart they are the same as testosterone; they bind to the same receptor, then enter the nucleus and, presumably, activate the same genes. It will be more interesting to investigate different non-steroid-based pathways for enhancing muscle growth. However, before we do that, we need to explore some other compounds linked with the steroid chemist Patrick Arnold. These are the hormone prodrugs.

Steroid prodrugs

Two molecules already featured in Figure 15—androstenedione and dehydroepiandrosterone (DHEA) are on the immediate pathway to anabolic steroid biosynthesis. They are therefore known as prodrugs. The idea is that inside the body the doped molecule will be converted into testosterone, though it is worth noting that they can also be converted into the female sex hormone estradiol, which may not be exactly what is desired.

DHEA is made in the adrenal glands. In the 1980s and 1990s DHEA was hyped with the same vigour that the drug companies of the 1930s used to try and sell testosterone. A list of disorders treatable by DHEA included an A–Z of illnesses from asthma to viral encephalitis via diabetes, malaria, and most things in-between. Like testosterone it was also assumed to have sports performance and anti-ageing benefits. The present-day scientific reality is, not surprisingly, more sober; DHEA can have genuine health benefits, but they are mostly restricted to disease conditions where the body does not make enough on its own, such as Addison's Disease where the adrenal glands are malfunctioning.

Androstenedione, perhaps because it is 'closer' to testosterone on the metabolic pathway (Figure 15) has come under more scrutiny by sportsmen. It was used first by the East German state, but acquired prominence in the West when Patrick Arnold imported it into the USA; it then became popular with top US sportsmen. Most famously Mark

McGwire, the baseball player, was found with a bottle of androstenedione in his changing locker in 1998, the same season he broke the home run record. At the time 'andro' use—although on the Olympic banned list—was not restricted by the major US professional sports. It is not altogether clear why McGwire bothered with a steroid prodrug at the time. He later admitted to taking the real thing (i.e. anabolic steroids) throughout this same record breaking 1998 season. Perhaps he followed the 'belt and braces' approach to doping.

But do these prodrugs really work to improve sports performance? The answer is almost certainly 'no' (chapter 5 ref. 7). Not only are the performance effects equivocal, but also there is no evidence that they can increase testosterone levels, at least in young people. This is not really surprising. The body has strict controls in place to maintain the levels of its hormones. If testosterone levels increase then mechanisms will kick into place to reduce its synthesis or enhance its breakdown. Taking an occasional, or even regular, prodrug is likely to have little effect unless it is combined with targeting, and inhibiting, these control mechanisms.

The one caveat is that, perhaps not surprisingly, most of the studies have been done in male athletes. Short-term andro studies on females have shown a rise in testosterone.[10] But for obvious ethical reasons this has not been followed through with longer-term controlled studies. My suspicion is though that in females, like males, the body's control systems would eventually reduce the levels of testosterone. After all testosterone is present in normal females, just at lower levels than males, so it is likely they have similar control mechanisms in place.

Steroid releasing agents

There are a whole class of compounds that, while not acting directly on muscle cells themselves, can work indirectly by triggering the release of other hormones. In principle these could be as effective, or even more so, than the cellular-targeting hormones themselves. Chief among these is the banned doping agent gonadotropin-releasing hormone (GnRH). This molecule is produced in the brain and, as its name implies, it controls the release of gonadotropins. Gonadotropins, such as Luteinizing Hormone (LH), travel from the brain and stimulate the gonads (the sex organs). The movement of hormones around the body can follow a complex pathway. GnRH is produced in the hypothalamus and it activates LH production in the pituitary gland; LH then activates testosterone production in the testes, which then, as we know, affects muscle cell growth.

LH activates testosterone production in both the male testes and the female ovaries. Despite this gonadotropins and their releasing factors are only banned in males. This does not necessarily mean they are not performance enhancing in females. However, doping detection in females would be very difficult. Changes in the menstrual cycle and pregnancy cause such large changes in the normal gonadotropin levels in females that it is difficult to define what constitutes an abnormal level. The most high profile athlete to have been accused of using gonadotropins is Manny Ramirez—another US baseball home-run slugger. In 2009 Ramirez was banned by Major League Baseball (MLB) for 50 games for taking a prohibited substance. Although the details of MLB findings are kept secret the substance was widely reported to be human chorionic gonadotropin.[11, 12] Ramirez himself merely stated that he had been given a banned drug by his doctor for unspecified personal health reasons.

It may seem strange to go to the high-tech expense of doping in this way, especially as there is no research to show the benefits of these testosterone-releasing compounds. If you are unsuccessful in raising testosterone levels you will have no performance benefit. If you are successful you will anyway test positive for testosterone in a drug test. Why not cut out the middle man and just take the risk of injecting testosterone in the first place? The answer probably lies in the dopers trying to subvert the body's control systems. Artificial anabolic steroids frequently down-regulate the production of the body's own testosterone. The body senses it has too many steroids and reduces its natural production to try to return to its previously balanced state. But this results in a lower than normal level of natural testosterone. By taking a testosterone-releasing compound at the same time as the artificial anabolic steroid, the athlete hopes to counteract this effect, gaining the simultaneous benefits of the normal and abnormal steroids. As the testosterone levels remain in the normal range this part of the process would, at least, be undetectable.

Human Growth Hormone

When people get old their muscle function declines. Every generation has its wonder anti-aging drug that attempts to put a stop to this process. Testosterone was the drug of choice in the 1950s and DHEA in the 1990s. Human Growth Hormone (HGH) is the first such drug of the twenty-first century. Typing 'human growth hormone health benefits' into a search engine will immediately give you a sense of the wide range of benefits claimed, from sexual potency to enhanced memory and vision. You won't be surprised to find out that there is no evidence to support these extraordinary claims. You also are probably only a bit surprised to feel that most web sites claiming to sell human growth hormone (either implicitly or explicitly) do nothing of the sort. Instead they sell some amino acids and herbs that contain minute amounts of compounds that are claimed to increase the release of the body's own growth hormone.

There are two reasons for taking this roundabout route. The first is that if you give almost homeopathic concentrations of a compound, but accompany it by marketing hype, you can maximise the placebo effect whilst minimising any drug side effects. But there is also a supply side issue. Real human growth hormone has historically been hard to come by. Unlike the hormone insulin, where diabetic patients can make use of insulin purified from the pancreas of slaughtered pigs, animal growth hormones do not work in humans. Therefore the only way to get hold of workable growth hormone was from the pituitary glands of dead humans. The pituitary gland is located at the base of the brain and is the size of a pea. Supply difficulties therefore meant that, unlike diabetics and insulin, people with growth deficiencies were restricted in their ability to access human growth hormone. Dead human bodies (from donors) were the only source. But there is a fate worse than growth deficiency. In 1985 cases of vCJD (variant Creuzfeldt-Jakob Disease)—the human form of 'mad cow disease'—were found in people who had received growth hormone produced from cadavers. The problem is that due to the small size of the pituitary gland and the difficulty in preparation of the hormone, many human brain samples were 'pooled' to create the product. This increases the chance of the eventual pharmaceutical containing the disease-causing agent.

So what was to be done? Enter probably the biggest success story of GM (genetic modification) technology. Insulin and human growth hormone are proteins. The DNA coding for these human proteins was extracted and transferred into *E. coli* bacteria—in this case a safe laboratory strain that doesn't cause food poisoning. Another piece of DNA was then stuck in front of the hormone gene. This second piece of DNA signalled for the gene to be continuously activated. This turned the bacteria into a factory dedicated to making one protein—a human hormone. As this process of genetic technology involves recombining (mixing) separate bits of DNA from humans and bacteria, the proteins produced are called recombinant proteins.

The terms rHGH or rEPO therefore refer to proteins produced in this way. However, the proteins themselves are not bacterial and human mixtures—they have identical amino acid compositions and function identically to the same proteins produced in a human. Now at last hormones could be made cheaper and safer—a real boon to people suffering from inherited growth hormone deficiencies.

Each generation's anti-ageing cures—testosterone, DHEA, HGH—have also been touted as a panacea for enhancing muscle strength in young people. So soon after the availability of rHGH in 1985, it rapidly became one of the drugs of choice in the doper's armoury. Athletes hope that HGH can deliver strength gains without the side effects of sex hormones. Growth hormone can increase the body's levels of IGF-1.

Dopers are seduced by the fact that the mTOR (GO) pathway in protein synthesis is activated by IGF-1 (as seen in the last chapter). However, there are problems with this idea. First the most dramatic effects of HGH on increasing IGF-1 are seen in people who are IGF-1 deficient in the first place. Secondly, and most importantly, remember that the mTOR GO pathway is not muscle-specific. It is no good inducing protein synthesis all over the body; you need to localise the response to muscle cells as much as possible.

There is some evidence that HGH can increase the collagen proteins necessary for strengthening tendons;[13] stronger tendons might well be useful to someone attempting to increase their muscle mass. However HGH on its own cannot increase muscle protein synthesis or strength.[14] Any weight gain appears to be mostly due to fluid retention. It is also not without other side effects including joint stiffness, muscle pain, and high blood pressure. So why do people keep taking it? I surmise two possibilities. The first relates to the fact that anything that can cost up to $20,000 a year on the black market has to be good for you; the athlete's version of shopping therapy. The second is that we are in the same place scientifically as we were with testosterone in the 1980s. Maybe just like then the doping coaches and athletes are right and HGH really works; we scientists have just not been clever enough to devise the proper ethical experiment to show the effect. My suspicion is that in this case the scientists are right and the dopers are wasting their money—though it has to be said I would not be completely surprised to be proved wrong.

Exotic drugs and the future of muscle doping

We have covered most of the drugs that we know have been used by athletes to improve their strength. But what does our knowledge of the pathway involved tells us about future developments? Genuinely new paradigm-shifting advances in sports performance drugs are likely to be related to compounds originating in studies in academic and pharmaceutical laboratories. Not surprisingly myostatin has come under intense scrutiny from the bodybuilding community. Just take a look again at the double muscled bull if you want to see the incentive. For surely if a genetically-induced reduction of myostatin led to massive muscle growth, then an athlete could get the same effect by taking a pill that inhibited myostatin?

There was even a candidate drug called follistatin, a natural protein that seemed to inhibit myostatin function. Increasing muscle mass could be of interest in a range of diseases where muscles become weakened, such as muscular dystrophy. For a while there was indeed interest from big pharmaceutical companies. Could a twenty-first century pharmaceutical race to produce myostatin inhibitors to cure muscle wasting diseases have the same effect on sport as the 1930s pharmaceutical race to create testosterone and its derivatives? Well maybe—the starting gun has been fired, but no one is yet anywhere near the finish line. For example follistatin itself binds to too many other things in the body to be useful as a drug when given externally; more success will likely require finding a way to increase the body's own follistatin production. However, the side effects may be severe—follistatin is present and active in every cell of the body, not just muscle cells, and we are not at all clear of all its normal physiological roles.

What about completely novel myostatin inhibitors? Wyeth Pharmaceuticals did make such a product, called myo-029, that underwent preliminary clinical trials for treatment of muscular dystrophy. Like many current drugs this is an artificial antibody. The body produces protein antibodies to bind tightly to invading bacteria and viruses and target them for destruction. The same properties can be used by pharmaceutical companies to create novel antibody molecules that can bind tightly to any target—in this case myostatin was the selected target. The idea was that myo-029 would zero in on myostatin in the blood and render it inactive.

However, the results were not good enough to warrant taking drug production further. Although the drug appeared safe, and some of the muscle fibres isolated from their body did seem to show improvements, none of the patients showed significant strength increases. Although

Wyeth subsequently claimed that they would look at other ways to block myostatin, their recent acquisition by the pharmaceutical giant Pfizer has probably put an end to that line of research; Pfizer have no interest in muscle wasting diseases. Curiously this one corporate pharmaceutical merger is likely to be more effective at slowing the pipeline of new muscle building drugs for athletes than any initiative by the anti-doping agencies.

Are there other places to look for the next generation of muscle building drugs? One ray of hope (or despair depending on your point of view) is how little we really know about the molecular details of the biochemical pathways we have been discussing. We know little about the proteins that modulate the testosterone receptor in the cell or the nature of the hormone response elements to which testosterone binds in the cell nucleus. There must be future opportunities for smart drugs to work at these sites. But progress could be slow. Though modern biology is very good at identifying proteins that can bind to other proteins, it is less good at proving the interaction is physiologically relevant. At the last count well over 100 proteins have been shown to have the ability to interact with the testosterone receptor.[15] But exactly how important these interactions are in the body is unclear. Understanding will require serious investment. If a deadly virus suddenly appeared that induced muscle wasting, drug companies would get involved and a myriad of new doping agents would appear. As it stands though diseases involving lack of muscle cell growth, such as muscular dystrophy, are fortunately rare and hence attract rather limited funding.

In contrast to muscle wasting, diseases involving uncontrolled cell growth are common. Pfizer's anti-cancer division was probably much larger than the whole output of Wyeth when they merged. As we noted before testosterone increases the size of prostate cancers. So there is a lot of money being spent trying to stop the cell growth functions of testosterone. Should this be successful it could have a significant impact on sports doping. At the minimum it will mean we understand more about how testosterone works at the molecular level. However, the effects could be more profound. Any therapy that stops testosterone working might suggest a suitable anti-therapy that would make testosterone more effective. This is already starting to happen. As we mentioned before GnRH increases the production of testosterone. GnRH works by binding to a receptor molecule. Molecules that stop GnRH binding to its receptor are therefore therapies for prostate cancer as they decrease the production of testosterone.[16]

In contrast a molecule that binds to the GnRH receptor and activates it is likely to increase testosterone production, at least in the short term. Literally thousands of these agonists have been synthesised in an effort to develop potent therapeutic compounds. Just as with the synthetic testosterone derivatives manufactured in the last century, these GnRH mimics are similar in structure and function to the natural hormone, but many times more potent. Any of these could enhance testosterone levels. Of course the flip side is that designing a drug that is the opposite of an anti-cancer drug could have carcinogenic properties itself. However, as we learnt before, an increased risk of developing cancer in the future is unlikely to put someone off doping in the present.

Our molecular analysis of the pathways that signal muscle cell growth has revealed that there is likely to be a significant range of molecules that can affect muscle mass. The most effective molecules are the anabolic androgenic steroids. They work in theory and in practice, especially in female athletes. Smaller effects can be seen following the use of creatine and, possibly, HMB. However, many other performance-enhancing drugs are nothing of the sort, both the biochemical theory and research evidence being absent. Nevertheless our limited knowledge of the mechanisms underpinning strength training, coupled to the vast number of potential undiscovered molecular targets, suggests that the details of this chapter will not stand the test of time.

7 Stimulants

'Woe to you my Princess, when I come, I will kiss you quite red and feed you till you are plump. And if you are forward, you shall see who is the stronger, a gentle little girl who doesn't eat enough or a big wild man who has cocaine in his body.'
Sigmund Freud (letter to his fiancée)

So far the drugs we have investigated in detail are those that have, by and large, been only of interest to sports people. But now we will discuss a class of performance-enhancing drug that is of much wider use in society. Current estimates suggest stimulant use at between eighty and ninety per cent of the population in most countries; indeed it is a rare person who has not tried one at some time in their lives. In fact I am taking a drink of just such a (legal) stimulant as I write these words. The links between social and sports drugs are therefore far more blurred than when we were talking about steroids or EPO. When it comes to stimulants many drugs are banned for athletic use primarily on the basis that they are illegal recreational drugs rather than that they are genuinely considered to be performance-enhancing.

When I started to write this book I was bit cynical when it came to the performance-enhancing effects of stimulants. I was reminded of the chapter in *Breaking the Chain* where Willy Voet was looking after the top French cyclist, Richard Virenque on the Tour de France (chapter 1 ref. 2). Immediately prior to a key time trial Virenque requested an injection of a 'special' concoction he had heard about from a rival team. Even Willy, no stranger to doping, was worried about using something he knew nothing about. However, he eventually relented and injected Virenque. Virenque rode the time trial of his life and responded: 'God, I felt good. That stuff's just amazing. We must get hold of it.' In truth Voet, mindful of injecting his elite charge with an unknown mixture an hour before one of the most important races of his life, had swapped the magic potion for a sugar solution. As Voet said: 'There is no substitute for self-belief. The bottom line was that there was no more effective drug for Richard than the public. A few injections of "allez Richard" going round his veins, a big hit of adoration to raise his pain threshold, a course of worship to make him feel invincible.'

This is the barrier we will come up against again and again in this chapter when trying to discriminate fact from fiction. Stimulants act on the brain to enhance performance.[1] Yet the brain is also the target of the placebo effect. If you think something will improve your performance, there is a fair chance that the thought itself will have a positive effect. It is therefore difficult to disentangle this placebo effect from a genuine effect of a stimulant. There are other issues; with stimulants we are dealing with a class of drugs that is designed to give us a 'high' to perform at our optimum. But elite sportspeople train to get the best out of their body at the critical moment. Why do they need a psychological enhancement? Even if they do how can we be sure that a stimulant is really having a direct effect? The Richard Vireneque experience tells us that, even more so than with other drugs, it is difficult to trust anecdotal tales from athletes themselves.

In clinical trials the placebo effect is controlled for using double blind trials; neither the patient nor the doctor know if they are getting the drug or a control. But even the 'gold-standard' randomised double-blind trial can have its problems if the volunteers can tell if they have been given a drug. Many clinical trials have been 'unblinded' due to the side effects of taking a drug being so obvious that the recipients know whether they are in the experimental control or the test placebo group. The same problem exists with a stimulant; subjects can potentially tell whether they have been given a stimulant or not by, for example, an increase in their heart beat. If they know they are on the stimulant they get a strong placebo effect—for surely a stimulant must improve performance? They think they can perform better and so they do.

A good example of this problem—in this case a placebo effect enhanced by a bit of wish fulfilment on behalf of the experimenter—can be seen in the first proper trials of a stimulant in exercise performance. Perhaps somewhat surprising our hero is none other than Sigmund Freud. Growing up with a mother steeped in behavioural psychology rather than psychoanalysis, I was always wary of Freud. This was not helped by his media persona in numerous TV programmes and films as a pompous pseudoscientific Austrian with a bad accent, championed by even more pompous psychiatrists such as TV's *Frasier.* So I was stunned to find that some of the very first quantitative sports science experiments were performed by Freud. Of course being Freud he chose a controversial drug, in this case cocaine. In fact cocaine was something of a passion for Freud—both scientifically *and* sexually as illustrated in the quotation that opened this chapter. In his scientific trial he noted the effects on his general perception: 'I took for the first time 0.05 g of cocaine and a few moments later I experienced a sudden exhilaration and feeling of ease'.[2]

But what about any ergogenic effects? In contrast to my stereotypical view of Freud he chose not to trust what his mind was telling him and decided with regard to these observations to 'render them objective through measurements'.[2] Sure enough in 1885 he published the first paper demonstrating increases in hand grip strength after cocaine administration.[3] Surely a promising career in academic sports psychophysiology was thrown away by his invention of psychoanalysis? Perhaps not—later studies have failed to reproduce the ergogenic effects of cocaine, at least not in fit and unfatigued athletes.

So what are stimulants and how might they be able to enhance sports performance?[1] Unlike steroids, stimulants are not defined by their structure, but their function. The scientific definition broadly agrees with what the average person might think of from everyday life. A stimulant is just an agent that makes something go faster—in the scientific sense this usually refers to the central nervous system. Therefore to understand how a stimulant works we need to know how signals in the brain are turned on and off.

The key cells in the nervous system are called neurones. The primary signal that activates them is electrical (see Figure 19a). But this signal is propagated by chemicals. An initial voltage change triggers a release of a neurotransmitter molecule from the end of a neurone. This neurotransmitter can then react with receptors on the surface of a nearby neurone. The signal can be positive (excitatory) or negative (inhibitory). If enough positive signals are sent this triggers an electrical signal from the second neurone. So the message is propagated. Multiple neurones make multiple connections with each other, so the pathways can get quite complex. However, the basic biochemistry is always the same.

The body produces hormones that can act on the brain, either directly as neurotransmitters themselves, or indirectly by modifying the action of other neurotransmitters. For example take adrenaline (known as epinephrine in the US). This is produced in the adrenal glands above the kidney. It acts on the central nervous system to activate pathways that can increase heart rate, increase the ability of the lungs to take in oxygen and increase blood flow throughout the body. Adrenaline is a general example of a compound the pharmaceutical companies love to mimic—compounds that 'piggyback' on the body's own control systems. The multibillion dollar industries in stimulants, depressants and antidepressants is a witness to the attempts to produce molecules that act on the central nervous system to turn on or—equally importantly—turn off signalling in the brain.

So how can a stimulant affect neurotransmission? Neurones are bathed in colourless cerebrospinal fluid (CSF) rather than blood. Therefore a stimulant has first to cross the blood-brain barrier that separates the CSF from the blood supply. This restricts stimulants to relatively small molecules. Once a stimulant reaches a neurone there are three ways it can increase the ability of the cell to stimulate its receptor (see Figure 19b). It can increase neurotransmitter release; it can prevent the re-uptake of the neurotransmitter back into the cell; or it can bypass the neurotransmitter altogether by directly binding to the receptor itself. Common clinical and recreational drugs work via all three mechanisms. Ecstasy increases neurotransmitter release, cocaine inhibits neurotransmitter uptake and caffeine directly acts on neurotransmitter receptors. Some drugs even work by multiple mechanisms: amphetamine increases neurotransmitter release at the same time as inhibiting neurotransmitter uptake, while ephedrine increases neurotransmitter release and directly stimulates receptors.

Things are not always as simple as this. Both cocaine and the antidepressant fluoxetine (trade name Prozac) inhibit neurotransmitter uptake. Yet they have very different physiological effects. The difference probably lies in the nature of the neurotransmitter being affected. Take three important neurotransmitters—dopamine, serotonin and noradrenaline (also known as norepinephrine). Both noradrenaline and cocaine can act on all three of these neurotransmitters whereas fluoxetine predominantly only affects serotonin. Stimulants such as methylphenidate fit somewhere between cocaine and fluoxetine. Methylphenidate can increase dopamine and noradrenaline in the brain, but has a limited effect on serotonin.

In principle you ought to be able to predict the action of a stimulant by knowing which combination of neurotransmitters it affects along with its mechanism of action. In practice we don't really know enough about the brain to be able to do this properly. Most drug companies still develop new stimulants largely by an expensive trial-and-error technique. Even if we could make accurate predictions for the human population, brains are complex organs and it would still prove difficult to predict the effects of drugs on a specific individual. Therefore when we study performance effects we will have the same problem as we have had before when trying to extrapolate from studies on patients, or healthy volunteers, to possible effects on elite athletes.

For an extreme example we just need to look at the history of the (in) famous stimulant methylphenidate. Discovered in 1944 it was identified as a stimulant in 1954. As a brain stimulant it is, perhaps not surprisingly, used to treat disorders such as narcolepsy, lethargy and depression. However, its claim to fame is largely via its use under the Ciba-Geigy trade name of Ritalin. Ritalin is famous for the calming influence it has on children with attention-deficit hyperactivity disorder (ADHD). A stimulant to calm people down? There is perhaps no clearer example of the effects of the subject on the effectiveness of a drug. It really matters who you are when you take methylphenidate. It is not exactly clear what is the reason for this, but one idea is that by increasing the levels of dopamine in the brain, methylphenidate increases the ability of children with ADHD to concentrate on specific tasks. Thus stimulation of a specific function in the brain leads to a change in *general* behaviour that would be labelled a decrease in stimulation by a layman.

However we should be cautious about simple extrapolations to sport. This is especially likely to be true of stimulants, which are generally (ab) used prior to the start of a race—arguably the moment where athletes are mentally most distinct from the general population. Elite athletes at the starting gate are hyperfocussed—perhaps the diametric opposite of a sufferer of ADHD.

There has been far more pharmaceutical money invested in stimulants and antidepressants than in anabolic steroids. But, in terms of sporting performance, research—and indeed widespread use by athletes—has overwhelmingly focused on two drugs, amphetamine and caffeine. Both have in their time been freely and legally available in society and sport. Now they sit on opposite sites of the regulatory fence in both spheres. A detailed comparison of the

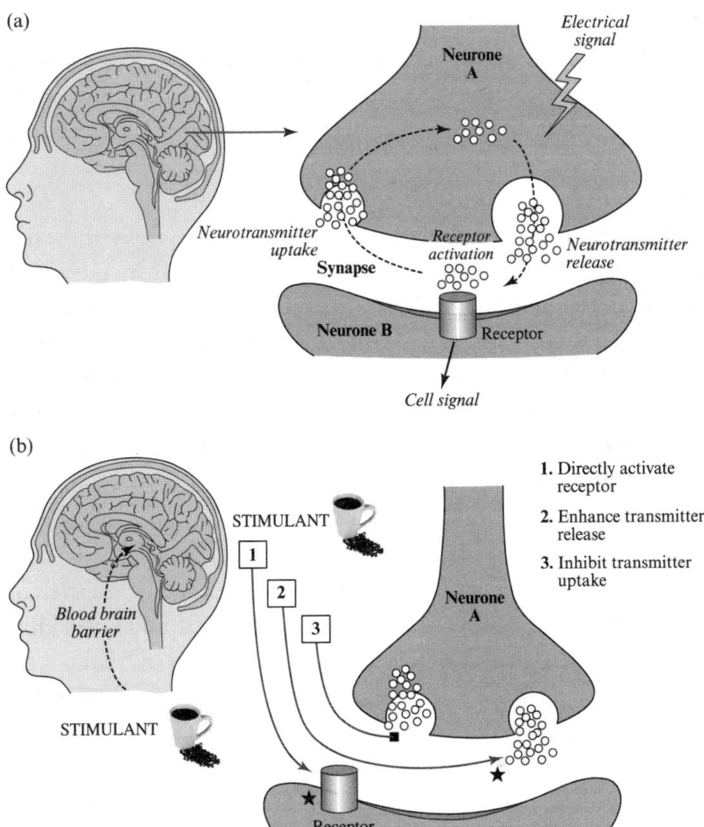

Figure 19 Stimulants and brain signalling

Neurotransmitters convert electrical messages into chemical signals between cells (a). The amount of signal formed is a balance between the release and uptake of neurotransmitters at the synapses between nerve cells (neurones). Brain stimulants (b) act to increase the activity of neurotransmitters by: 1 *directly activating receptors i.e. mimicking the action of the neurotransmitter;* 2 *enhancing the release of neurotransmitters i.e. enabling more signal to be formed;* 3 *inhibiting the uptake of neurotransmitters i.e. allowing the signal to last longer* mechanism and ergogenic properties of amphetamine and caffeine should therefore result in a general understanding of the benefits, pitfalls and future possibilities that stimulants can bring to sport (chapter 3 ref. 14).

Amphetamines

Amphetamine is an example of a hormone that can act directly on neurotransmitter release. Many recreational drugs work by minor alterations of the amphetamine molecule and so would be expected to have similar effects. The dominant effect of amphetamine is to increase the levels of the neurotransmitters noradrenaline and dopamine. This has a range of physiological consequences (see Figure 20). Noradrenaline triggers the 'fight or flight' response, resulting in increased heart rate and increased blood flow to muscle. In principle this response could be performance enhancing. Dopamine, on the other hand, can control motor function in the brain as evidenced by Parkinson's Disease sufferers who have lowered dopamine levels. So there is potentially a direct effect of amphetamines on the control of motor function in the brain (via dopamine) as well as the secondary effects in the heart and muscle (via noradrenaline).

However, dopamine has subtler mood altering effects that can be just as important as its effects on motor control. Dopamine acts on a region of the brain called the mesolimbic pathway.[4] This is responsible for 'rewarding' the brain when it performs well. Activating this pathway is likely to be responsible for the short-term mood-enhancing effects of amphetamines. Whether these mood effects actually act to improve performance is a separate question.

Any performance-enhancing effects of amphetamine are likely to be a consequence of the activation of adrenaline and dopamine receptors (Figure 20). These receptors themselves are therefore also potential targets for doping agents.

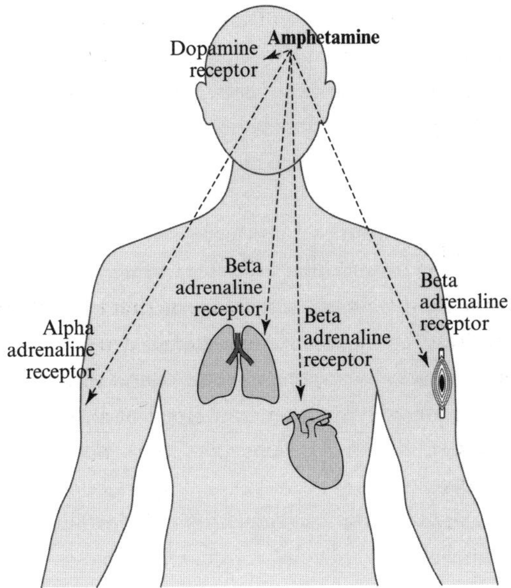

Figure 20 Amphetamine

Pathways of amphetamine action in the brain (via release of dopamine) and other organs (via release of adrenaline).

ADRENALINE RECEPTORS

Noradrenaline—and adrenaline—act predominantly on receptors called, unsurprisingly, adrenoreceptors. Different receptors exist in different tissues. In total there are nine different types of adrenoreceptors. Each of these initiates signalling phosphorylation cascades. As in the case of muscle growth the opportunities for developing drugs to inhibit and activate these downstream pathways are, if not endless, at least seriously underdeveloped. And there is commercial interest. Adrenoreceptors are part of a class of molecules called G-protein coupled receptors. A massive forty per cent of current prescription drugs directly target these kind of receptors. As for the remaining sixty per cent, several of those target downstream phosphorylation and dephosphorylation reactions.

There are two main types of adrenoreceptors. The first type—called alpha adrenoreceptors—predominates in the skin. Activating these restricts blood flow. You don't need a high blood flow to the skin to run away from the sabre tooth tiger; better to conserve your energy where it really matters like the heart, leg and arm muscles. However, activating these peripheral receptors still has its uses in today's generally more sedate society. Drugs that activate alpha receptors are able to restrict peripheral blood flow and hence decrease fluid loss from the body. Vasoconstriction of the blood vessels in the nose, throat, and sinuses, is exactly what you need when your nose swells and you are producing copious fluid and mucus. Alpha receptor activators are designed to relieve these symptoms.

This is why stimulants such as ephedrine, pseudoephedrine, phenylephrine and phenylpropanolamine have been used in over-the-counter cold medications. In general stimulants such as these can be considered the poor man's amphetamine. They are less able to cross the blood brain barrier and have limited interactions with dopamine metabolism. So they have fewer central effects on the brain—it is difficult to get a 'rush' from taking an over-the-counter cold medication. Even worse, restricting skin blood flow can have unfortunate consequences in sport. When racing in hot climates good skin blood flow is needed to cool the blood—and hence the body. Stimulation of the alpha receptors can therefore cause overheating. It seems possible that this was a contributory factor in the most (in)famous case of amphetamine abuse—the death of the English cyclist Tommy Simpson who collapsed on the climb up Mont Ventoux in the 1967 Tour de France.

The second types of adrenoreceptors are called beta receptors. There is a variety of drugs available that can activate these receptors directly.[5] These include anti-asthmatic agents such as Ventolin (salbutamol), which are designed to act on adrenoreceoptors in the lung and open the bronchial airways. Salbutamol is not banned provided an asthmatic athlete claims a medical exemption. However, suspicions of misuse persist, in part due to the number of medical exemptions requested. When directly compared to general populations recent studies suggest that the proportion of elite athletes diagnosed with asthma is not different from the general population.[6] Yet one in five US athletes who competed in the 1996 Athens Olympics reported in a questionnaire that they had been treated for, or had symptoms consistent with asthma; as many as ten per cent of athletes at the Sydney Olympics four years later has medical exemptions allowing them to take inhaled medication. In some sports the prevalence of asthma symptoms amongst elite athletes seems too high to be due to chance (over sixty per cent of cross country skiers and forty per cent of long distance cyclists in the US Olympic teams in the 1990s).[7] Exercise can induce asthma-like symptoms and it may well be this that causes the very high prevalence of reported breathing difficulties. However, suspicions abound that, at least historically, some of this reported high incidence was due to an attempt by these athletes to get their hands (or rather lungs) on salbutamol.

Nevertheless it seems that asthma inhalers are unlikely to be restricted in sport. If nothing else it is difficult to perceive what real performance benefit can ensue in a healthy individual. What is much more concerning to anti-doping agencies, however, is when beta receptor activators are taken in a high dose pill form, rather than as an inhaler. Receptors are present in the heart and skeletal muscle. So at high doses these drugs can directly increase heart rate and muscle blood flow. One such orally active drug, clenbuterol also appears to have anabolic (muscle building) as well as anti-asthmatic effects. In this case unlike most anti-asthma medication, it is completely banned for use in sport, just as if it were an anabolic steroid.

Drugs can be designed to turn off signals as well as turn them on. Beta blockers are a class of drugs that inhibits the action of beta adrenogenic receptors; they are especially effective in the heart.[5] They are widely used to treat heart failure, high blood pressure, heart arrhythmia and angina. However, they are also on the banned list in sports. This ban was introduced following infamous incidents such as when in 1984 the entire Italian Olympic shooting team claimed a medical exemption for the use of beta blockers. Either there was an amazing correlation of heart failure and the ability to shoot straight, or else the team was trying to gain an unfair advantage.

Why would such an anti-stimulant be performance-enhancing? The key is in the ability to control heart beat and tremors. This makes beta blockers potentially performance-enhancing in sports such as pistol shooting. Acute control of your body is the key to hitting the target. Elite athletes even train to squeeze the trigger on the pause between their heart beats. Beta blockers may be a useful adjunct here. It is also possible that dropping your heart beat quickly could be useful in the biathlon. In this winter sport you have to ski a 5 km circuit and then shoot at five targets. You lose time for each target missed. It may be beneficial to be able to drop your heart beat quickly so that you can shoot better after skiing. But the problem is that what you gain in shooting you are likely to lose in the subsequent skiing as the beta blocker slows you down.

Sometimes the desire to prevent people getting advantages by taking beta blockers seems to go too far. The Olympic sport of modern pentathlon traditionally consisted of fencing, shooting, horse riding, swimming and running. Taking a beta blocker for the shooting event theoretically meant that it had time to leave the athlete's body and not inhibit the next day's running performance. Therefore the order of events was changed so that the running event was made sooner after the shooting in order to prevent an athlete taking advantage of the delay in this way. Whether this was ever a problem is not clear; even without the change of timing an athlete could counter the problem by using a drug that is removed from the body more quickly. This whole process seems like a lot of fuss to solve a non-existent problem. To be effective a beta blocker would have to be taken during the actual Olympic competition itself; any winner using beta blockers could therefore almost guarantee themselves a positive drug test.

A sadder story resulted when snooker banned beta blockers as part of its attempt to get recognition as an Olympic sport. In fact all it succeeded in doing was ruining the career of an individual. The Canadian snooker player Bill Werbeniuk had a hereditary health condition that affected his hand—familial benign essential tremor. In order to control his cue action he needed huge amounts of alcohol, as much as twenty pints a day during the world championships. He replaced some of this with beta blockers on his doctor's advice, but was forced to give these up when snooker moved to accept the Olympic anti-doping list.

In shooting and snooker there is a strong rationale as to how beta blockers could be performance enhancing, even if in Werbeniuk's case they are probably better described as performance-enabling. The justification seems much less valid in sports such as ski-jumping, bobsleigh and luge. The rationale for the ban here seems to be that they could reduce anxiety and make it easier for people to compete well in these high-risk dangerous sports. I can understand why I would need beta blockers (and probably alcohol) to be persuaded to hurtle down a mountain on a glorified tea tray, but I am somewhat less convinced that it would make a difference to a skeleton luge expert such as the 2010 UK Olympic gold medal winner, Amy Williams.

DOPAMINE RECEPTORS

Amphetamine can act via enhancing dopamine production in the brain. What about other molecules that act in a similar way? Dopamine—and compounds that are precursors like L-DOPA—are predominantly used to treat diseases where the body's own dopamine levels are decreased, Parkinson's Disease being the most common. However, given dopamine's critical role in brain function there is a raft of compounds that have been developed to modify its effects. These include dopamine receptor activators, dopamine receptor inhibitors, dopamine transport inhibitors, dopamine transport activators and dopamine activity enhancers. Most of these have found niche roles in treating certain brain dysfunctions or as recreational drugs. As yet there is limited evidence of them being abused in sport, or even if they would be effective. The few compounds that have been tested are those that are used clinically already—in particular the dopamine uptake inhibitors used to treat ADHD such as methylphenidate. However, these have not been shown to have an enhancing effect on sports performance under normal conditions (chapter 2 ref. 3).

CAN AMPHETAMINES IMPROVE SPORTS PERFORMANCE?

You know when you have taken amphetamines. Of that there is no doubt. There is no shortage of athletes feeling that they can perform better when 'charged up'. The cyclist Paul Kimmage describes the feeling the first time he takes amphetamine in his book 'Rough Ride' (p. 146).

> Personally I feel the effort. I feel the effort, the shortness of breath on the climb, but mentally I am so strong that it doesn't matter. My mind has been stimulated. Stimulated by amphetamines. I believe I am invincible therefore I am.

However, in the same book we find that he is forced to admit that 'a lot of it is purely psychological' (p. 169). Deciding to race drug-free he finished eighth in a race when he felt he was 'riding on the same level as everyone else'. He later finds out this is one of the races where everyone uses amphetamines. Racing clean again the next year he doesn't even bother—he knows he can't win and finishes well down the field.

Of course calling something 'purely psychological' is a loose way to put it scientifically speaking. As a biochemist I am comfortable with the notion that psychology works via modifying chemicals in the brain. This does not mean that we can 'explain' psychology in terms of chemicals. But it does mean that anything that has a measurable psychological effect will come with an accompanying baggage of neurochemical changes. So dismissing something as purely psychological begs the question. Taking stimulants and thinking you have taken stimulants will both affect the balance of key chemicals in the brain.

The key question is do the stimulants enable you to reach a performance peak that is unobtainable by drug-free means? The scientific evidence is equivocal. In a 2002 review,[9] Ron Bouchard summarised thirteen studies on the use of amphetamine or related compounds in laboratory tests of athletic performance, including swimming, cycling, running and throwing. In only four out of thirteen studies were positive effects seen. There was also no obvious link between the strength of the dose, the type of activity and whether or not amphetamines improve performance. For example running and swimming were both part of studies showing changes that were significant and others where no effect was seen.

In terms of understanding performance effects nothing much has changed since 2002. I could as easily cite a recent paper showing a positive effect as one showing none. The evidence base is certainly far weaker than with regards to other drugs we have discussed such as EPO and anabolic steroids. If there is a direct ergogenic effect, independent of the placebo effect, it must be small. It seems likely that, if they work at all, amphetamines are acting on conscious control processes in the body. A recent study proposed just such a mechanism.[10] Elite cyclists were asked to cycle at an effort between 'hard' and 'very hard' for as long as they could. Those on the amphetamines 'chose' to cycle at a harder pace even though they did not perceive it as such. Taking amphetamines appeared to overcome a control mechanism in the brain that limited performance.

If this idea is true then understanding the possible benefits of amphetamines on athletes will require an intimate knowledge of the mental limitations of performance. Why do we slow down and what makes us choose to stop running? Is our decision to stop running really a conscious decision, or is it one that we later interpret as having been conscious? This story reminds me of a statement I made to a keen research student who was studying the effect of the perception of effort on exercise performance. Although his experimental manipulations used psychological rather than pharmacological tricks, the problem was similar to that I discussed above. I couldn't see how he could address his research questions without knowing how consciousness worked. He didn't seem to think this was too much of a problem, though I suggested it might be a bit ambitious for a three-year project ....

In science you occasionally come across these roadblocks. In this case—as I advised the student—you have to remember Sir Peter Medawer's old adage. If politics is the art of the possible, then science is the art of the soluble.[11] By this he meant that you sometimes need to find the soft underbelly of a scientific problem rather than tackling it head on. For consciousness research this has recently meant brain imaging, whether by positron emission tomography (PET) or more usually functional Magnetic Resonance Imaging (fMRI). These tools can image the brain blood flow changes that accompany the activation of specific brain regions. Unfortunately both these techniques require a static head and a body encased in a large magnet. It is hard to imagine a method less cumbersome for studying exercise, especially when the goal is to work your whole body to exhaustion. So whilst there are hundreds of fMRI studies looking at the effects of amphetamine abuse on brain function, there are no studies looking at the acute effects of amphetamine on exercise.

One solution is to look at non-invasive brain imaging techniques that are more portable. My own research looks at literally shining light on the brain to look at changes in blood flow in response to neuronal activation.[12] This enables me to investigate brain function even when people are moving about. However, even this technique is difficult to apply in intense exercise as subtle blood flow changes due to the activation of brain pathways can be masked by global effects. In particular there are dramatic changes in oxygen consumption and carbon dioxide production during exercise that can impact directly on brain blood flow, masking any subtle effects due to conscious manipulation of the brain and the perception of fatigue.

However, until these problems are solved, I am stuck with my own prejudices—or what I prefer to call my scientific intuition—in reviewing an admittedly incomplete scientific literature. My personal view is not too different to that I phrased at the start of the chapter. Amphetamines have the potential to exert powerful effects on the body's chemistry and physiology. But any benefit is unlikely to improve the performance of an athlete who is at the peak of their preparation for a one-off event.

The situation changes over the long term; athletes cannot always perform at their peak. A psychological let down is common. For example, I watched the 2010 London Grand Prix athletics event. This was supposed to be a glorious homecoming for the most successful British athletics team ever to perform in the European championships. The Grand Prix was two weeks after the championships. The winners received the adulation of the crowd; and then promptly proceeded to lose all their events. This is after two weeks of rest in events many of which require little recovery time. The fatigue must surely have been psychological, rather than physical. These are the sorts of conditions of mental fatigue when stimulants such as amphetamines could be effective. In this case it is possible that a pharmacological stimulus could overcome the psychological fatigue.

Now I am not suggesting for a moment that these particular athletes would be tempted to take drugs. But there are sporting events that require athletes to perform continually at their peak and require considerable mental as well as physical effort. The Tour de France requires cyclists to perform at—or close to—their peak on a daily basis for three weeks. I am therefore not surprised that amphetamines were de rigueur on the Tour de France before drug tests were introduced.

Caffeine

When you look at the effects of stimulant drugs on sports activity, one molecule stands out from the field. There are more, and better controlled, studies that show the positive effects of caffeine than any other stimulant.[13] There are the usual caveats in that many of these have not been with elite athletes. But as caffeine was only ever banned during a competition itself, it has even been possible to do research studies with elite sports people in the off season. Now that caffeine is not banned at all, I would be amazed if caffeine supplementation did not form a major part of many athlete's pre-race preparations. It can improve any event that requires exercise for over five minutes (the effects on shorter distances are more controversial). And when it comes to running or cycling, caffeine can increase the time to exhaustion by as much as twenty to fifty per cent.

So how does this wonder drug work? Caffeine works on so many processes in the test tube that it is very difficult to decide what are the really important effects in the body. As well as modulating neurotransmitters caffeine can have direct actions outside the brain. For example it can both perturb metabolism and alter muscle signalling.

The key question is which mechanism is relevant in the body after an athlete drinks a couple of double espressos or other high caffeine drink? Figure 21 indicates all the possible effects of caffeine that have been identified in the test tube. Some mechanisms can be ruled out as they require doses of caffeine requiring a continuous intravenous drip of double espressos—even then if the requisite levels were reached they would probably be toxic. Although it might be possible to design a drug that could activate this chemistry at lower non-toxic concentrations, such a 'designer caffeine' would surely be immediately banned by the anti-doping authorities as an unnatural enhancement to performance.

This does rather beg the question why caffeine is not banned. Indeed at first caffeine use was made a doping offence, though somewhat bizarrely the levels that resulted in a ban were higher than those considered to be performance enhancing. However, since 2004 this ban has been lifted; it is now fine to have a coffee before you run—at least if you are a human, it is still banned for racehorses. The rationale here, one assumes, is that caffeine is considered a normal part of a human diet, but horses don't usually hang out at Starbucks. Recently, the introduction of caffeine pills by athletes has led to concerns about whether to reintroduce a ban. Although these pills are no more effective than drinking two or three very strong espressos at raising caffeine levels, the negative perception of taking a pill seems to be the concern. Speaking in July 2010, John Fahey, the President of the World Anti-Doping Agency said that taking pills is against the 'spirit of sport'.[14]

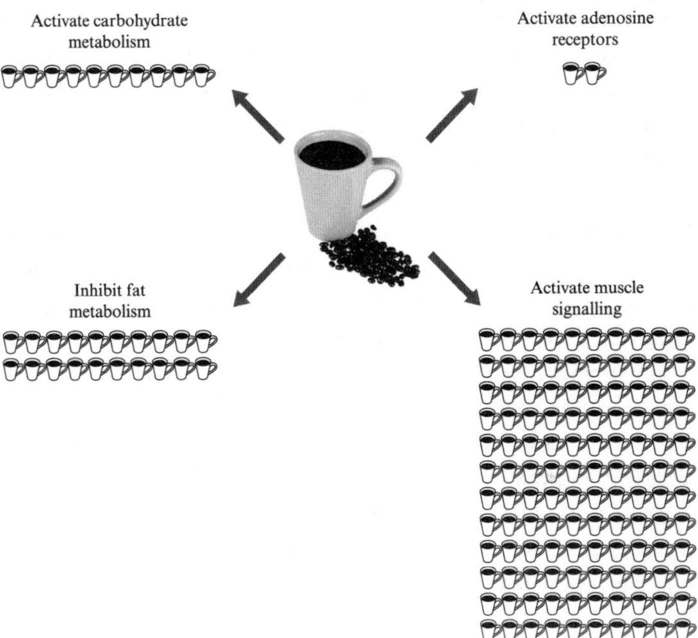

Figure 21 Coffee and metabolism

Illustration of the dose of caffeine required to perturb different biochemical pathways in the body. (assuming one cup of coffee contains 0.1 gram of caffeine resulting in a concentration in the blood of 2 milligrams per litre).

Data taken from Figure 4 of Jones, G., (2008) 'Caffeine and other sympathomimetic stimulants: modes of action and effects on sports performance' Essays in Biochemistry 44, 109–123.

So how does caffeine work and why is it so much better than amphetamines at enhancing performance? There is a growing consensus that its ability to act at low concentrations as a brain stimulant is the key (chapter 3 ref. 14). In contrast to amphetamines, caffeine does not act to affect adrenaline levels. Instead it affects a different messenger molecule called adenosine. Caffeine does this by binding to the receptor in the brain designed for adenosine; this binding occurs at caffeine concentrations that are easily accessible following a couple of your favourite caffeine-based beverages. Normally adenosine binding to these receptors causes an inhibitory effect on neurotransmitter release. However, when caffeine binds, adenosine cannot bind. The inhibitory receptor remains inactive. Two negatives make a positive; the inhibition of an inhibitory receptor increases the release of neurotransmitter. Caffeine has stimulated the neural pathway.

The most important effect of caffeine seems restricted to a specific subtype of adenosine receptor in a region of the brain called the basal ganglia. This region influences the control of motor functions linking emotional states and physical activity. Just like the yin yang STOP GO messages in muscle synthesis there are two pathways in this region of the brain. One of them stimulates motor activity and the other inhibits it. Dopamine is the key neurotransmitter in these pathways. By binding to receptors in the inhibitory pathway it turns off the pathway. Whilst adenosine would normally work to limit dopamine function, adenosine cannot function in the presence of caffeine. Caffeine and dopamine, working in concert, acts to switch off the pathways that limit motor activation signals in the brain. This is the most likely reason for the performance enhancement.

To summarise: the inhibition of an inhibitor leads to the activation of an inhibitor of an inhibitory pathway. This is the point where most people might be tempted to give up on biochemistry! But if we think of ourselves as a dopamine molecule it can help to understand the process involved (see Figure 22). Our job is to keep the muscle working. The system used is analogous to the dead man's handle on a train; the driver needs to keep holding the handle down or a brake is engaged and the train stops. Imagine that as dopamine we are like the driver's hand keeping pressure on the handle. But the train (body) cannot go on forever. At some time it gets a signal to stop. How does this work? Normally an adenosine signal tells the brain the body is tired. It does this by removing dopamine, thus releasing the handle. Enter caffeine. Caffeine stops adenosine working. This allows us to keep our hand on the handle. So the brake stays unengaged and we can exercise forever—or actually till the next protective mechanism kicks in, but hopefully this is only after we have won our gold medal.

The exact details of the downstream signalling in this caffeine modulating motor control region is unclear. However, it appears to involve a phosphorylation cascade. A key initial protein in this pathway goes by the catchy moniker of 'dopamine- and cyclic AMP-regulated phosphoprotein with a molecular weight of 32'—fortunately known to its friends as DARPP-32. Funding for this research is likely to be forthcoming. Not because of sports doping, but rather to address issues relating to recreational drug use. For whereas caffeine activates DARPP-32 in the basal ganglia, opiates such as heroin activate DARPP-32 in regions of the brain linked to the pathology of drug addiction; an area of obvious interest to medical researchers.[15]

Understanding pathways at this level of detail gives us more than an insight into the wonders of biochemical pathways and control. It illustrates once again that the compounds that enhance performance—and caffeine is undoubtedly one of these—are only the tip of the iceberg of the possible range of compounds that can be envisioned. Inhibitors of DARPP-32 phosphorylation may stop drug addiction—they may also increase sports performance. Adenosine receptor antagonists are pharmacological gold dust. The specific receptor type caffeine activates (called A2a) is implicated in conditions such as Parkinson's Disease, anxiety, mood disorders and even memory loss in Alzheimer's Disease. Indeed the link between the A2a receptor and impaired memory in Alzheimer's has led to findings that support the use of caffeine to partially reverse this memory loss.[16] This is likely to lead to an exponential increase in the already large number of adenosine A2a receptor antagonists available. I would not be surprised if sports dopers were not eyeing this emerging pharmaceutical market with enthusiasm.

I have lingering thoughts that, as with amphetamine, the positive effects of caffeine could be irrelevant for an athlete in an optimum psychological state of preparation. However, I am less convinced in this case. Caffeine turns off an inhibitory pathway. In general turning off inhibitory pathways seems to be the best way to prolong or enhance an activity without compromising the integrity of its function. Viagra works in the same way—it turns off an inhibitory pathway and prolongs

The dopamine sensitive "dead man's switch" blocks the inhibitory pathway and enables exercise

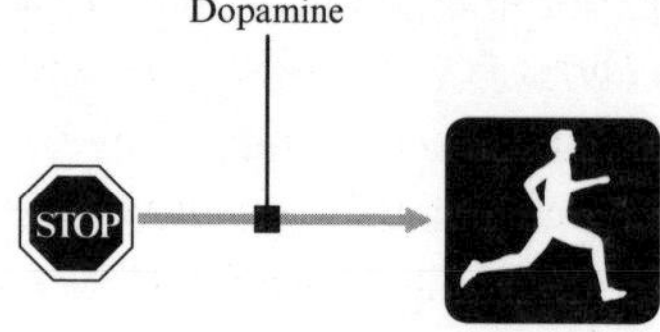

Adenosine stops exercise by removing dopamine and enabling the inhibitory pathway

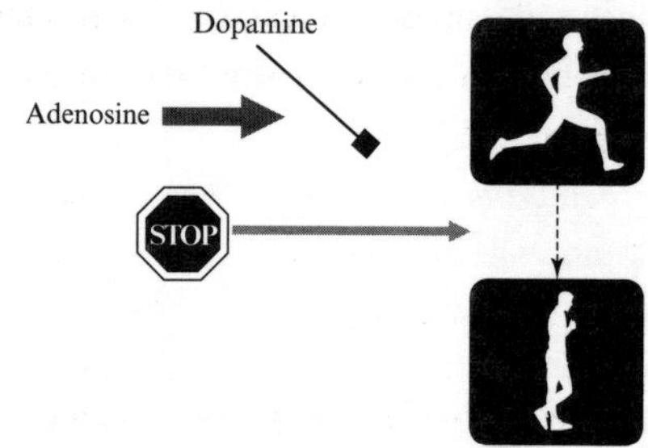

Caffeine prevents the inhibitory brake being initiated by removing adenosine

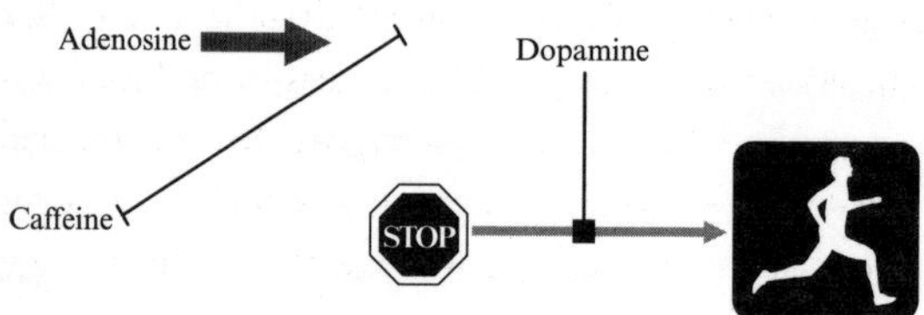

Figure 22 How caffeine keeps you running

'My enemy's enemy is my friend'. Caffeine and dopamine act synergistically. Caffeine stops adenosine stopping dopamine stopping the inhibitory pathway that stops exercise.

sexual activation. I am also minded of the dramatic effects resulting from inhibiting the muscle OFF pathway—remember the double muscled bulls with decreased myostatin?

From our detailed comparison of caffeine and amphetamine we can make some broad conclusions. Amphetamines can induce a range of system-wide effects, any and all of which might have a small performance effect in an athlete, especially if they are already fatigued at the start of exercise. Direct action on the brain seems diffuse and untargeted, mostly affecting mood resulting in an unclear effect on sports performance. However, although caffeine targets the same neurotransmitter in the brain—dopamine—its effects are targeted to the specific region in the brain that controls motor function. This seems to give a far greater performance benefit than amphetamine's more general effect. Take amphetamines and you feel you can run faster; take caffeine and you actually will run faster. Put crudely amphetamine talks the talk, but caffeine walks the walk.

Other banned stimulants

There has been a range of other banned stimulants suggested to enhance performance if taken before an event. As discussed earlier in this chapter the three decongestants pseudoephedrine, phenylephrine and phenylpropanolamine work via their effects on alpha adrenoreceptors. But they can have some effects on heart rate, blood pressure and metabolism. Athletes have attempted to take them in higher than the normal clinical dose in order to take advantage of these secondary effects. Of the three only pseudoephedrine remains in general use, there being concerns over side effects of using ephedrine and phenylpropanolamine. In the USA it is difficult to get any of these compounds, even pseudoephedrine, in high doses, as they can be used to manufacture illegal drugs such as amphetamine. This is problematic for the average person hoping to get a performance enhancement; the latest study suggesting that an average 80 kg male would need to take a 200 mg dose for optimal effect. Even in the more liberal UK, this is one third of the pills in the largest packet size you can buy.

Stimulants of this kind have been the cause of many of the most controversial disqualifications, with athletes—or even worse their coaches—not checking whether they should take a specific medication during a competition. Ephedrine almost lost Linford Christie his silver medal in the 1988 Olympics for example. However, the most notorious case was that of the sixteen-year-old Rumanian gymnast Andreaa Răducan. At the 2000 Sydney Olympics the Rumanian women's gymnastics team obtained a clean sweep of medals in the overall event, with Răducan taking the gold. However, Răducan subsequently tested positive for pseudoephedrine and had her gold medal withdrawn by the International Olympic Committee. The drugs test result was a consequence of treatment for a cold that had struck many members of the Rumanian team. The team doctor had prescribed the painkiller Nurofen, which in its form as a cold medication also contains pseudoephedrine. There was a minimum level of pseudoephedrine you were allowed in your urine drugs test. This is not corrected for body mass. Răducan being the lightest gymnast registered over the limit; her team mate Simona Amanar, who finished second, also tested positive for pseudoephedrine, but at below the banned level. Amanar was given the gold medal, but boycotted the medal ceremony claiming that Răducan was the true champion.

We can safely assume that Răducan was not trying to gain a performance advantage by taking her Nurofen pill. But others clearly have tried this trick. Pseudoephedrine was removed from the doping banned list in 2003. It was re-banned in 2010 as more and more athletes were found to have high levels in their urine. Interestingly there may be a lesson for society at large here. At least in sport, 'legalising' a specific drug clearly increases the use of that drug. And banning it decreases use. The threat of losing your gold medal appears to be a real disincentive. But is there really a benefit to be had from taking pseudoephedrine? The results from scientific studies are a bit more promising than amphetamine, but not by much. Part of the problem is that many of the most positive studies have been in combination with caffeine, so it is difficult to be specific about which precisely is the beneficial agent, or indeed whether there is a synergistic combination effect. Nevertheless there is some evidence that ephedrine can increase muscle endurance in the short term and that pseudoephedrine can improve power output in short term cycling and reduce the time taken to run a mile (chapter 3 ref 14).

Improving the mind

The stimulants discussed so far relate to sports performance of a particular type. Can I run faster, can I perform longer without getting fatigued? Apart from the case of pistol shooting and beta-blockers they don't relate to enhancing the skill of the performance itself. But improving your mental and physical skills is exactly the area where society is currently targeting its stimulant use. In the 1980s a top lawyer might be taking cocaine to enable them to party longer, nowadays they are as likely to be taking modafinil to make them work smarter. Even academics are not immune. In June 2007 the *Times Higher Education* supplement published an article entitled 'Pills provide brain boost for academics'.[17] The claim was that academics—and students—were using so-called 'smart' drugs to improve their performance. Of course there is history here. Caffeine has long been one of the student drugs of choice—increased doses are taken at critical times. But this is mostly to enable an 'all-nighter' to beat an essay deadline rather than to intrinsically improve the quality of the essay itself.

The first smart drugs seemed to be acting in a similar way to caffeine, in that they enabled activity at times when the body was telling someone to sleep. The exemplar is the drug, modafinil. This is clinically approved to treat sleep disorders such as narcolepsy. It is also occasionally prescribed in diseases where fatigue is a major symptom, such as multiple sclerosis. It is of interest to the military as it is claimed to enable long periods of sleep deprivation with limited side effects. In recent years, however, modafinil use has mushroomed in the general population. This is partly as a result of 'off label' prescribing. This is when doctors prescribe a drug licensed for one condition to treat another; they can do this if it is in their clinical judgement in the patient's best interest. This is how modafinil is used in multiple sclerosis. However, drug companies are not allowed to market such untested therapies.

In the case of modafinil the manufacturers had not exactly gone out of their way to discourage uses outside the area where they had clinical approval. In 2002 Cephalon Inc. pleaded guilty to promoting the use of modafinil inappropriately.[18] Modafinil was approved by the US Food and Drug Administration 'to treat excessive daytime sleepiness associated with narcolepsy' but Cephalon labelled it 'as a daytime stimulant to treat sleepiness, tiredness, decreased activity, lack of energy and fatigue' i.e. not just for the small number of patients with narcolepsy-induced fatigue. The settlement agreement for this – and other drugs mislabelled by Cephalon – cost the company well over $400 million.

But how do modafinil and other smart drugs work? One supposed smart drug we met earlier, methylphenidate, works by increasing dopamine and noradrenaline in the brain. But modafinil has an altogether more mysterious method of action.[19] It has been implicated in increasing the levels of a wide range of neurotransmitters. This does not just include those we have discussed earlier such as noradrenaline, serotonin and dopamine, but also others we haven't met such as histamine, glutamate, GABA (gamma-aminobutyric acid) and orexin. It has even been suggested to work via a direct effect on the energy producing organelles, the mitochondria. In short the field is a mess. Anyone who thinks they understand the situation—drug marketers included—simply doesn't know the facts. My best guess is that either there is a complex interplay caused by small increases in multiple brain transmitter pathways, or there is a key research finding missing.

The 'hottest' class of brain enhancers go much further than targeting sleep deprivation. They claim to improve cognitive function in Alzheimer's and Parkinson's Disease. Called ampakines they work by enhancing the abilities of the neurotransmitter glutamate to activate receptors in the brain implicated in memory and learning. Glutamate in high doses is a brain toxin; so it is clearly better to enhance its 'memory' function by making it more effective rather than by making more of it. Hence the interest by drug companies.

It is worth pausing here to note the sheer magnitude of candidate drugs that can be identified by modern pharmaceutical screening. A protein called *CREB* has been shown to affect the transition from short term to long-term memory. A relatively simple chemical screen detected 1,800 compounds that activated this gene and could—at least theoretically—enhance memory function.[20] This illustrates the opportunities modern drug discovery programs can offer to unscrupulous dopers.

However, as any pharmaceutical company executive will tell you, it is a long and expensive (≈$1 billion) road from identifying a potential compound to producing a workable and safe drug therapy. So how close are we to this goal with memory enhancers and smart drugs and can we enhance the general healthy population? The popular press claim that we are on the cusp of a revolution in brain function, heralded by the breakthroughs of methylphenidate, modafinil and ampakines. Maybe there is something in Cephalon's illegal marketing strategy after all? Just as an aspirin can cure our headache, a dose of methylphenidate will make even your 'normal' kid smarter in school while modafinil will get them into a top university. Yet as we have learnt throughout this book improving top performance is a very different story to removing an obstacle to performance.

Modafinil and methylphenidate are clearly useful tools for people who are clinically fatigued or hyperactive. But in the general population their effects are far less than the hype. A 2010 analysis of all the scientific research suggested rather limited effects.[21] Methylphendiate did have minor memory enhancing effects and modafinil was successful in maintaining wakefulness in well-rested individuals. But more grandiose 'brain enhancing' effects were not found. Indeed modafinil was unsuccessful in maintaining brain performance over a long period of use—the subjects felt they could overcome their sleepiness, but in fact still performed badly, a degree of overconfidence that might be disturbing if used in a military context. Ampakines are even more experimental; at present, there is no evidence that these compounds work in patients with compromised function, let alone that they would work to enhance healthy brains.

Things are no better with 'natural' brain enhancers. The signature stimulant 'pill' in the health food store—Gingko biloba—has also taken a knock recently. A 2007 review of the scientific literature[22] concluded that there was 'no convincing evidence from randomised clinical trials for a robust positive effect of G. biloba ingestion upon any aspect of cognitive function in healthy young people, after either acute or longer term administration'.

In short we are not looking at a new breed of drug-enhanced superhumans, nor should universities rush to introduce a drug testing regime in their examination halls anytime soon. But what about sports? Smart drugs are banned only in so much as they are listed as artificial stimulants and hence banned during competition. In theory this would allow an athlete to use a smart drug to improve performance in learning a task in training that later helped his performance. In terms of documented use as far as we know modafinil is the only smart drug that has been used. Earlier we heard about the apparent epidemic of heart disease in Italian pistol shooters. In 2003, a similar wave of narcolepsy seemed to hit athletics. The most famous case was the US sprinter Kelli White. She tested positive for modafinil after winning the 100 m and 200 m titles at the 2003 World Athletic Championships in Paris. At first she claimed narcolepsy was the reason she took the modafinil. However, most observers thought that this was just an excuse for attempting to gain the benefits of using an illegal stimulant. These suspicions turned out to be correct when White admitted to the US Anti Doping Agency (USADA) that she had indeed taken a range of performance enhancing drugs during this period. In 2004 USADA banned her for two years and she was stripped of her 2003 gold medals.

Why did White risk her career by taking modafinil? There is no real evidence that modafinil is any more or less effective than any other stimulant, and certainly it would seem to be an unlikely drug to use in a sprint event. The likely reason it is chosen is a 'belt and braces' approach by the dopers. It is cheap, has limited side effects and can be used outside competition. Even in competition, at the time it was assumed to be undetectable. So you may as well add it to the mix if you are taking steroids and EPO anyway (as Kelli White was) and be hung for the proverbial sheep as a lamb. White was linked to the same BALCO company that supplied other athletes. Indeed Victor Conte prescribed modafinil to the British athlete Dwaine Chambers as a 'wakefulness promoting' agent before competitions in order to decrease fatigue and enhance mental alertness and reaction time in the 100 m sprint (chapter 1 ref. 9). This seemed a huge risk given that to be effective it had to be given one hour before competition and so would be liable for detection in drug tests.

In a handful of sports cognitive ability of the type university students try to enhance with smart drugs is relevant. Orienteering, or example, requires an ability to read maps when physically exhausted. However, there is little money or fame—at least outside its Scandinavian heartlands—in orienteering. In other sports athletes and coaches are unlikely to be swayed that such woolly concepts as 'wakefulness promoting' are likely to be performance enhancing. Anyone who sees a sprint race can be in no doubt that the athletes are awake and alert. A 100 m sprinter spends most of his time trying to calm his nerves, not psych himself up. However, if a drug could enhance motor control and improve reaction time, the fastest man out of the blocks or the top marksman could be drug-enhanced. I think this is not an impossible dream—or nightmare depending on one's view—but we are not there yet by a long way.

Concluding words

One final point about stimulants must be emphasised. With regards to their use, both in sports and everyday life, whilst a little of what you fancy does you good, more is not always better. With all molecules that bind to receptors and initiate a signalling cascade there is a law of diminishing returns. The body adapts to a new situation and downregulates. That is why in the scientific studies, the experimental subjects always have to withdraw from caffeine use at least a day before the test. It is also why many studies show no difference in everyday performance and alertness of people who take caffeine versus those who don't. The difference is all in the change. Taking caffeine for the first time will give you a buzz and make you alert. Going off caffeine will make you drowsy (and probably give you a withdrawal headache). But it is pretty much impossible to tell apart people in everyday life who take caffeine versus those who don't. The same is probably true for sports performance; this is one of the reasons stimulants are only banned in the races themselves and there are no sanctions against their use in training.

Earlier in this chapter we learnt of the death of Tommy Simpson from heat exhaustion on an alpine pass. We should perhaps end on a similar cautionary tale. What effects could result from a successful stimulant that overcame the psychological limits to performance? Some sports psychologists and physiologists take the extreme view that these controls are designed, consciously or unconsciously, to stop the body undergoing catastrophic systems failure. Most take a less apocalyptic position. But all view with trepidation anything that interferes with the brain's control mechanisms.

Take a recent example. Drugs that inhibit dopamine and noradrenaline uptake have been shown to have genuine performance benefits (chapter 2 ref. 3), but only if the exercise takes place at high temperature (the two temperatures compared were 18°C (64°F) and 30°C (86°F)). The core body temperature of those on the drug increased along with their performance, reaching levels in excess of 40°C (104°F). But there were no perceptual differences. Those experiencing the higher temperatures on the drug felt no greater discomfort than when they were exercising without the drug—they just exercised longer and harder. One of the drugs tested was methylphenidate.[23] You can't help feeling that, given the number of legal child prescriptions for methylphenidate and the covert adult self-doping, another Tommy Simpson accident might be on the cards—history repeating itself, but with a different mechanism.

In conclusion we can draw a useful distinction by thinking of two different classes of stimulants. On the one hand we have those like amphetamine that are linked to recreational drugs, tend to affect our mood, and give us the perception that we can run, throw and swim better than ever before. This is probably what most people think of as a stimulant. Although these drugs do seem to be effective in enhancing the performance of athletes when they suffer from day in day out fatigue, in most cases they are unlikely to affect an individual record performance. And if they worked better, they probably would be dangerous as they would turn off the body's natural defences. On the other hand we have stimulants like caffeine, which have a rather specific effect on the brain and consequent exercise performance that is independent of their social use. How caffeine makes a student run faster does not share a common mechanism with how caffeine keeps the same student awake to write their all night essay (or a professor to write a book chapter).

In the next chapter we will see a field that is even more hyped than stimulants. In this case we know, however, that there are massive performance increases possible. The Dachshund and the Greyhound share most of the same essential genes. But one can run a bit faster. Can we mimic these breeders and engineer our genes to be more Black Beauty and less Muffin the mule?

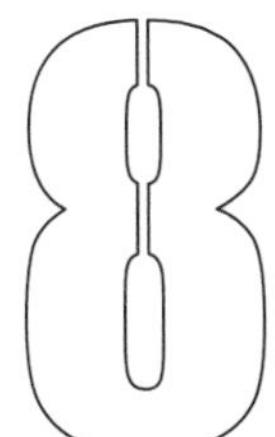

Gene Doping

'We were bred for it.... We were simply bred for physical qualities.'

Lee Evans, 400 m winner at the 1968 Olympics, *Sports Illustrated*, 18 Jan., 1971.

Figure 23 Evans is pictured in the middle waving a Black Panther beret during the medal ceremony. (c) AP/Press Association Images

A question of breeding

Genes matter. In most things in life if you want to succeed a key component of success is to pick the right parents. Musical talent, health, longevity—all these have a strong genetic component. It would be amazing if sporting achievement did not also obey these same rules. Dennis and Leslie Compton in cricket (and football), Venus and Serena Williams in tennis, Ralf and Michael Schumacher in Formula 1 motor racing are all surely testament to the importance of the right genetic inheritance. But these athletes all shared the same environment and this matters too. Your genes interact with your environment from your first hours in the womb to the last minutes in the nursing home. It is not good enough to simply state the famous sporting relations who became elite athletes as evidence for a strong genetic component in sport.

It is notoriously difficult to separate the genetic and environmental components of success. You might think identical twins would provide the answer. However, the key studies—studying twins separated at birth and hence reared in different environments—have focused on psychological and health traits, rather than sporting ones. In any case controlling an environment completely is impossible. Take Eric and Ted Bedser. These identical twins were considered inseparable throughout their lives. They dressed identically, played cricket and football together in the same teams, were educated at the same school and made their professional cricket debuts in the same match in 1939. At one time they were both the elite of the game: fast bowlers who could terrorise batsmen with fearsome deliveries. However, when they joined Surrey cricket club they decided they needed to diversify to be successful in the same team. A coin toss ensued. Eric lost the toss. He had to switch from aggressive fast to slow spin bowling. This could have turned out all right—after all just as a slow toss from a wily knuckleball pitcher can defeat the most eagle-eyed baseball slugger, so many famous cricketers resort to the spinning trickery of a chinaman, googly or doosra delivery. However, it so happens that the Surrey cricket team at the time had two of the most famous spin bowlers of all time in Jim Laker and Tony Lock. Given limited bowling opportunities, Eric was forced to concentrate on his batting. His fate was sealed in his England Trial; Jim Laker played for the opposition. Laker was such a success that no one batsmen on Eric's team had a chance—in fact Laker had the unheard of bowling figures of eight wickets for only two runs scored, a cricket feat that even made it into *The New York Times.*[1]

Humbled in his trial, Eric never played in a competitive match for England. Alex on the other hand played fifty-one matches for his country, with his twin faithfully accompanying him on his overseas tours. This story illustrates two things: the capricious nature of the sporting environment our genes swim in, but—equally crucially—the fact that *our genes can create our environment.* The Bedser twins created a situation where their environments differed with profound effects on their ultimate sporting success.

There are ways other than twin studies of measuring the genetic contribution to sporting performance. These show that the physiological underpinnings of sporting performance such as oxygen consumption, cardiac output, muscle fibre type are no less heritable than other human characteristics. The figures vary from as low as twenty per cent nature and eighty per cent nurture to traits that seem almost entirely genetically determined.[2] The exact details depend on the method of analysis, but in general 'strength' characteristics seem more inherited than 'endurance' ones. While the latter are more susceptible to training, the nature of your muscles is much more controlled by the accidents of your birth. The ratio of fast twitch (sprint) to slow twitch (endurance) muscle fibres, for example, is under strong genetic control.

Can we alter genes to enhance sporting performance? To answer this question, we first need to control the environment as closely as possible. Fortunately as scientists we don't need to plan our own experiments as animal breeders have done much of the hard work for us. We need look no further than the world of thoroughbred horse racing. I can remember as a young boy thinking it unfair that you could only enter a horse in a race if you knew its whole family history and it came from approved stock. To my untutored mind this seemed outrageously unfair to the poor wild horse. Surely some poor Black Beauty was being deprived of the chance to win The Epsom Derby or the Breeder's Cup?

This was probably a pipe dream. As Charles Darwin states in the opening chapter of the *Origin of Species*[3] while 'Some effect may be attributed to the direct and definite action of the external conditions of life, and some to habit; but he would be a bold man who would account by such agencies for the differences between a dray and race-horse, a greyhound and bloodhound, a carrier and tumbler pigeon.' Darwin recognised that artificial selection had led the English racehorse 'to surpass in fleetness and size the parent Arabs'. The fine details of racehorse training may be a dark unscientific art, but at its core the selection of the right stock is still paramount. There is a reason why breeders paid $1 million for the Canadian racehorse *Northern Dancer* to 'cover' their mares. *Northern Dancer* sired a record number of Breeder's Cup winners and his bloodline has coursed through every Prix de l'Arc de Triomphe winner since 1994. *Sea the Stars,* arguably the best racehorse of all time, was his great-great grandson.

Of course things are not that simple. Even with thoroughbred horses, breeders know that running fast is not 100 per cent inherited. The top racer may not have the top genes; you cannot guarantee that the best horse will sire a winner every time. *Northern Dancer's* stud fee rose to such astronomical levels only when it became clear that his offspring were also winning races. *Sea the Stars* had a much more prestigious career than his great-great grandfather. Still in 2010 breeders could get in early with their mares at a bargain price of $100,000, only ten per cent of the fee that his great-great granddad could command twenty-five years earlier.

There are questions as to how far such a process of artificial breeding can be taken. It is possible that with racehorses and greyhounds we may be entering a situation of diminishing returns. The best horse now is indeed faster than the wild horse it was bred from, but since the 1920s the top race times have hardly changed.[4] *Northern Dancer*'s offspring are a small fluctuation on an essentially optimised machine. Breeders are stuck with the gene stock from the first wild Araby horses bred in the 1700s. Traditional horse—and dog—breeding techniques cannot select a completely different species. A Pekingese dog is still basically a wolf, although it does seem to be wearing sheep's clothing.

Still in sport we are not interested in creating a new species—just in becoming stronger, faster members of the human race. In this context you might assume that we have some way to go to mimic what has been achieved in animal sports. After all we can't ethically breed humans for speed; even if we could, the long generation time would mean that the 200 years it took to optimise a racehorse would take closer to 2,000 in humans. In the absence of such a breeding programme we must surely be a long way off our top performance? Surprisingly an interesting statistical analysis by Mark Denny of Stanford University in 2008 indicates that we may in fact be quite close to our maximum speed.[5] With some exceptions (the men's 100 m, the women's marathon) mathematical models do indeed suggest that we are pretty much at our limit of running performance. How can this be if we have not taken advantage of selective breeding? The answer almost certainly lies in the size of the genetic material available for selection.

All today's racehorses were bred from under thirty individual horses; ninety-five per cent of today's racehorses are descended from a single stallion. Contrast this with seven billion humans, many of whom crave the money and fame of an Olympic gold medal. A fast individual who appeared out of this diverse human gene pool could undoubtedly be improved by a small percentage by a genetic breeding programme. However, our population increase, combined with the worldwide interest in sport, is making the gene pool for selection of athletes very diverse. Human athletes are close to being as genetically optimised as the best racehorse. A theoretical world government trying to create the fastest human being in the shortest possible time would probably be advised to introduce a worldwide screening system for fast children rather than just breeding from the current crop of top athletes.

A question of race

Of course if we have a population of seven billion, it does not mean that a top runner is equally likely to appear from all parts of this diverse gene pool. Any discussion of genes and sport is going to have to address at some time the headline quotation to this chapter. Are the top sprinters always going to be black? Are the best long distance runners always going to be East African? Before entering this sociobiological minefield we need to unpick the question in the first place. We live after all in a deeply confused world when it comes to talking about race. Colour is everything. So Christophe Lemaitre can be hailed in a 2010 French press release as the first white person to run under ten seconds for the 100 m. Yet Patrick Johnson ran much faster than Lemaitre the year before. Johnson is half Irish and half Australian aborigine. He looks dark but genetically he is, if anything, even more distinct from black sprinters than Lemaitre.

There is a notorious 1994 book, *The Bell Curve*, co-written by the psychologist Richard Herrnstein and the political scientist, Charles Murray.[6] It argues that there are strong racial components to intelligence. Its sporting equivalent was penned by the journalist Jon Entine in 2004 and entitled *Taboo: why black athletes dominate sports and why we're afraid to talk about it.*[7] As expected this is a contentious idea. The concept of race—and hence the term black itself—is notoriously difficult to define biologically. There are powerful economic reasons why a social group that is economically deprived will look to areas such as sport for its validation, especially when sporting excellence is frequently accompanied by a significant economic benefit. For every racial argument there is a sociological counter-argument. We shouldn't forget the significant effects of role models. In Britain, our TV screens in sport are filled with successful black athletes but, as we all know, this effect diminishes when we start looking at our business and political leaders.

Society's preconceptions can have direct impacts on performance. Populations frequently perform to the prejudices we have of them. Take an elegant 1999 study by Jeff Stone at the University of Arizona.[8]

He gave black and white students a standardised task involving putting a golf ball. The subjects were told that the test was either a measure of 'natural athletic ability' or one of 'sports intelligence'. Although the task was completely identical, students performed differently depending upon the skill they thought the task required. Black students performed better when they thought the task was testing their 'natural athletic ability to perform complex tasks that require hand–eye coordination'. White students on the other hand were best when thinking they were putting on display their 'ability to think strategically during an athletic performance'. The placebo effect rides again; if we think we will do well, we will.

In 1995 Sir Roger Bannister—he of the sub-four-minute mile—stated that 'black sprinters and black athletes in general all seem to have certain natural anatomical advantages'.[9] But black athletes don't dominate all sport. Swimming is a case in point. Is this because they are less buoyant as has been suggested by some? Maybe so, but take the USA. The number of African Americans who can swim at all—let alone compete at an elite level—is far below the average in the general population. Cultural barriers limit the pool of performers and it is just as likely that this is as important as any direct genetically controlled anatomical difference.

We should be aware that it is only too easy to make generalisations from observing sports events. We all love to be the opinionated fan, but this can cloud our judgement of any underpinning science. In 1980 Michelle Platini captained one of the best football teams of its generation, winning the European Championship for France. The team was exclusively white. Yet almost twenty years later, the 1998 French team that won the World Cup consisted largely of players whose immediate forefathers had migrated from Armenia, Senegal, Ghana, Guyana, Argentina, Algeria and Portugal. The famous 'blanc, black, beur', ('white, black, arab') team swept all before it and followed their World Cup win with the European Championship two years later. Fast forward six more years to the 2006 World Cup final. Now the French fielded a team without a white player. Was this a sign of black genetic superiority coming to the fore? Was top flight football going to become an all black sport as Roger Bannister and Jon Entine might argue?

The problem is that you can make the completely opposite case just by slightly changing your time frame. The non-white 2006 French team was in fact beaten in the final by an all white Italian team; four years later Spain won the World Cup with arguably the best football team of all time. Like Italy, Spain did not field a non-white player. Yet it would be as absurd to argue from the evidence of the last ten years that you need to be white to win a football World Cup as it would be to argue from the previous ten years that black athletes were about to take over football.

Football is a complex sport. Maybe the genetic superiority is masked by factors such as coaching and tactics? Can a better argument be made by studying track and field athletics? Take Usain Bolt with his tall, muscular yet coordinated stature. Is this frame a genetic freak of nature? Does his success come from his black genes? Before jumping to these conclusions we shouldn't forget that, whilst some consider Usain Bolt to be a genetic outlier, his 2009 world record 100 m time was only 1.1 per cent faster than his nearest rival. But in the same year Paula Radcliffe's 2003 women's marathon world record still stood over 2.5 per cent faster than the second best time. It would be hard to find a whiter looking athlete than Cheshire-born Paula Radcliffe. Yet no one is arguing that white women are genetically advantaged in ultra-long distance races. And for sure no one is looking for a genetic hot spot in the North of England responsible for breeding elite female marathon runners. We fare no better if we look at boxing. Since Cassius Clay (Muhammad Ali) won the world heavyweight title in the 1960s it seemed to be a preserve of 'black' athletes. Yet since 2005 the division has been dominated by white Eastern Europeans.

So simple racial generalisations are deeply problematic. Instead we should look at the science—and by that I mean look at the individual's genes not their skin colour. There are of course some genes that control skin colour. But these may not be linked closely to those that control performance. I think the emotive aspects (and terminology) of race need to be set to one side. Ultimately we will need to go down to the level of the individual gene. Here racial stereotypes vanish. No one can be emotive when talking about something called an *ACTN3* gene polymorphism after all.

So what does the science tell us? The casual observer of the effect of genes on sports performance will usually make two observations that are 'obviously' true. Black athletes of West African origin are good at sprinting and those of East African origin are good at long distance running. This is 'obvious' because the 100 m and 200 m races are always contested by runners originally native to West Africa, whereas the 500 m, 10,000 m and marathon are always won by people from the eastern side of the continent.

However, closer examination reveals that, not only does this ad hominem have a problem with the racial classification of 'black' that we discussed earlier, but also that it has a problem with the geography. This is easiest to see when it comes to East African runners. For it isn't 'East Africa' that is special—it is two countries, Kenya and Ethiopia. Even in these two countries the top runners come from severely restricted areas. Over eighty per cent of the Kenyan athletes come from the Rift Valley, whereas in Ethiopia elite athletes predominate in the Arsi region that occupies less than five per cent of the country. In fact you could drop a blanket over two regions comprising less than one per cent of the African land mass and population and cover the birth place of over fifty per cent of Africa's top distance runners. Not surprisingly scientists have noted this and zeroed in on this group.

The case of Kenya is particularly intriguing as one tribe—the Kalenjin—comprises seventy-five per cent of their elite long distance runners; in 1988 one subset of this tribe—the Nandi—supplied 42.1 per cent of Kenya's elite runners from a base of only 1.8 per cent of Kenya's total population. Surely studying these geographically isolated groups would show they contained unique genes that enabled excellence in distance running? Or so the theory went.

Recently the genetic science has started to deliver. One of the key studies on individual elite athletes was undertaken by the University of Glasgow's International Centre for East African Running Science. In the case of the Ethiopians they showed that—as far as we can tell with current techniques—it is impossible to tell the elite athletes apart genetically from the rest of the population.[10] Even worse for those hoping to discover the magic of East African genes, some of the elite athletes shared more recent common ancestry with Europeans than with each other.

While these clusters of excellence were not genetically isolated, there was differentiation in cultural and environmental factors. Rural children, though genetically identical to those living in the urban environment, led a far more active lifestyle; it was common practice to run long distances to school each day. Of all the Kenyan tribes, the Nandi resisted urbanisation the most. The only sport they have is running and the rewards of running success are great. Where the cultural drive is weaker the resulting success is weaker; this may explain why female East African runners are still somewhat less dominant than their male counterparts.

Disentangling these cultural and biological phenomena is a complex business and only recently have interdisciplinary collaborations between sociologists, geneticists and physiologists started to bear fruit. It is probably too premature to be definitive, but the initial conclusions suggest there may be nothing intrinsically special about East Africans. Their population pool probably contains a higher percentage of people with the basic genetic requirements to be elite distance runners. When this is combined with altitude training from birth, an active lifestyle, strong cultural and economic rewards for success, positive role models and active peer group competition, you have a recipe for domination on the world stage. Similar arguments about gene–environment interactions can explain the current Jamaican dominance of sprinting.

The world of running has been here before. At the turn of the twentieth century Scandinavian runners dominated all distance events. For a brief exciting part of my life in the 1980s British runners dominated middle distance running. At the turn of the twenty-first century the mantle of track dominance has passed to East African and Jamaican runners. It is possible that this phase too will pass and a more heterogeneous pool of runners in both sprinting and long distance running will re-emerge, particularly if sociological changes in Africa and the Americas start to yield alternative rewards for young men and women.

Only one sport is truly global with mass participation in essentially every country in the world. That sport is football. The black-white-black-white alternation of success that is currently seen with football may well become the norm for all sports where the will to succeed and the rewards of success are equally open to all countries and cultures.

Are there specific genes associated with human performance?

Even though race may be a red herring when it comes to sports performance, there is no doubting that genetics plays a part. The question for dopers is 'what part does it play and can it be manipulated artificially?' If there is a limit to what natural breeding and gene pool selection will yield in terms of performance enhancement, can unnatural genetic selection do better? Can gene doping work?

We first need to know what specific genes can affect performance. This is an area where science has made rapid progress in recent years. The earliest genetic studies of exercise were funded for medical research. In 1992, the US National Institute of Health funded the HERITAGE family study. This explored the effects of genes on exercise in 650 Canadian and US subjects. The intention was to look at genetic and non-genetic factors that affected the response to aerobic exercise training.[11] What has this got to do with health? Exercise was known to be of benefit in combating heart disease. But it was also known that people responded differently to the same exercise therapy. Maybe if the genetic basis of this was known then therapies could be individualised to optimise the health benefit? Again sports performance research benefitted from a research programme aimed at health. For the HERITAGE study showed that a number of specific genes were responsive to exercise and could predict training performance. Now with the advent of cheaper genetic testing it has become possible to move on from studying the health of the general population to studying what underpins the performance of individual elite athletes.

What does it mean to look at a genetic profile? We are different because our genes are different. We might be 99.9 per cent identical to another human. But that 0.1 per cent creates the potential for far more than a 0.1 per cent change in function. The average human contains 25,000 genes on their DNA. Each gene codes for a protein that performs a useful function. Each protein is made up of different amino acids. Let's assume that a gene codes for about 500 amino acids. Then a 0.1 per cent variation could result in as many as 0.1 per cent x 500 x 25,000 possibilities; on average this would result in 12,500 amino acid differences between the functional genes of two people. These 12,500 changes can mix together in unpredictable ways. The number of different combinations that 12,500 numbers can be put together is termed its factorial, illustrated mathematically as 12,500 with an exclamation mark after it or 12,500! It is calculated by multiplying 12,500 x 12,499 x 12,498 x 12,497 etc. all the way down to 1. If you want to calculate 12,500! you can't use your normal calculator; you have to log onto a specialist web page.[12] To put the resultant answer in context a billion is one followed by nine zeros. The most famous large number—a googol—is 1 followed by 100 zeros. 12,500! dwarfs a googol, coming in at 45,786 digits long. This is why you don't run into an identical twin by accident; the only way for someone to have the same genes as you is if you once shared a zygote with them.

The situation gets even more complicated when we consider that most of our DNA does not directly code for a gene that makes a protein. However, this DNA can still indirectly affect exercise performance by controlling the activity of those genes that do make proteins. Adding together this useful non-coding DNA with the genes that make proteins, it has been estimated that there are 10 million possible variants in the human population—as many as 100,000 of which are likely to have some functional significance. The number of possible different combinations of these variants is a number half a million digits long; this would take up far more than the rest of the space in this book. At this point we might realistically throw our arms in the air and feel that there is no way we can determine the genetic effects on exercise. But there is hope. While it is possible that some genes work better in specific combinations, it is unlikely that they all do. So to a first approximation we can focus on the 100,000 functional mutations individually. Geneticists can then use selection techniques to explore which of these might be important in exercise.

How does this genetic variation express itself? A mutation is a one-off change in your DNA. However, we tend not to think of ourselves as being a bundle of mutations, mutations being such an ugly pejorative term. Instead the word scientists use to describe the changes in DNA that are present in natural populations—and make us who we are as individuals—is polymorphism. Just like proteins are made up of amino acids, your DNA is made up of individual nucleotides. As Watson and Crick—the discovers of the DNA double helix structure—showed, it is the order of these nucleotides that is key to our genetic code. So the simplest change between you and me would be a single nucleotide polymorphism. This is abbreviated to SNP, pronounced 'snip'. To a first approximation I am different to you because my snips are different to yours.

Population geneticists spend their life studying these SNPs. The basic strategy employed is called genetic association analysis. The method was developed by medical researchers looking for genes associated with diseases. The simplest association is a tight link between a mutation and a disease. The paradigmatic example is sickle cell anaemia. Here a single mutation on the oxygen transport protein haemoglobin causes the red blood cell to change from a healthy disk to a sickle shape, making it harder to pass through the small capillaries in the blood circulation. This might seem to be only bad news. However, it is harder for the malarial parasite to attack these sickle shaped cells. Thus the sickle cell mutation has health advantages and disadvantages—you may struggle to deliver oxygen at times, but you are safe from malaria. The result is that the mutant protein is favoured in parts of the world where malaria is endemic.

The genetic association in the sickle cell case is 100 per cent. If you have the sickle polymorphism, you have the disease. Other genes have a lower association with disease. If you have the polymorphism you have an increased chance of suffering, but you are not guaranteed to succumb. The most famous examples of these are the breast cancer susceptibility genes—*BRCA1* and *BRCA2*. The normal role of the *BRCA* gene is to repair the DNA in your body; this plays a role in preventing cancer. However, the presence of the polymorphisms *BRCA1* and *BRCA2* is associated with an increase in the likelihood of getting breast cancer. The association is such that the lifetime risk of getting breast cancer rises from twelve per cent to sixty per cent.

The genetic associations that relate to exercise appear to be like *BRCA* rather then sickle cell. At first glance there seems to be no one magic 'gene' that means you have superhuman strength or endurance. However, instead there is an increasing number of performance enhancing polymorphisms that improve your chances of doing well.[13] Over 200 of these have been identified.[14] However, it is not as simple as that. Just because a gene is associated with improved performance it does not mean that the gene itself causes the performance increase. One possibility is that the gene is close to another one on the chromosome and so tends to be inherited along with it. This second 'linked' gene could be the real reason behind the performance enhancement. So although elite athletes might all have a specific gene, adding more copies of this gene might have no effect (as it is the presence of the linked gene that really matters). A gene doper could waste a lot of their time going down this route. So for that matter could scientists trying to understand exercise.

The same issues occur in medical research. Therefore when a new polymorphism is associated with a disease the next step is to try and determine if there is a causal relationship. The best proof comes from replacing a normal polymorphism with the one that is associated with the disease. You can then test if this induces the disease. Of course even if this process were possible in humans it would not be ethical—instead cell lines or animals are used. Genetic studies of disease (usually in mice) comprise the vast majority of animal procedures carried out in the world. It is work of this type that showed a causal link between *BRCA* polymorphisms and cancer.

There have been very few, if any, such definitive studies to confirm polymorphisms that affect human performance. There are two reasons for this. First as we have discussed before there is a lot more money in medical research than sports research, but there is also an ethical question. To perform an animal study you need to convince an ethical committee that the procedure has the chance to benefit animal or human health. Understanding what makes people run faster would not pass this ethical test.

Fortunately there are other ways of closing the gap between mere genetic association and causal link. As well as being able to measure a person's complete DNA content (genome) it is now possible to measure which genes are being used at any one time to make new proteins. If a gene has a polymorphism associated with improved performance *and* is shown to be active during exercise, the case for a causal link is strengthened. This requires taking a piece of muscle (a biopsy). The first studies were medical; they were aimed at trying to understand why some people did not improve their performance during exercise therapy. Subsequently sports scientists showed that genes that are associated with a performance benefit are actually used during that exercise.[15] However, studies on elite athletes are still not possible. They are generally not happy with donating even a tiny bit of their precious muscle for a biopsy—at least until they have finished sport and are in retirement.

What are the genes that enhance performance?

Although it is useful to know that a specific gene affects performance is far less interesting than *why* it affects performance. Knowledge of a possible mechanism is an important extra piece of the scientific jigsaw. If a gene is associated with high performance *and* you can see *how* it could work, you are far more confident that you are onto the real deal, not some associated unimportant gene.

A list of all performance-enhancing polymorphisms would read like a phone book of acronyms. It is more fruitful to look at a few key genes. First up is *ACE*; a wonderful sounding acronym for a gene linked to exercise. In 1998 *ACE* was the first gene to be shown to have polymorphisms related to exercise performance in the classic paper called 'Gene for Human Performance' by Hugh Montgomery and co-workers at University College London.[16] The study used new army recruits undergoing basic training. In all human studies it is important to control the external environment as much as possible. The beauty of studying this group is that for ten weeks they live in the same environment, eat the same food and do the same training programme. This would be very difficult to organise outside in the 'real' world.

Rather than being a single change in a piece of DNA coding for an amino acid, *ACE* polymorphisms relate to the insertion or deletion of whole chunks of DNA. An insertion that makes the gene larger is called the 'I' form. The form without the insertion is called the 'D' form. As you get one gene from each parent, people can be classified into three categories—II, ID or DD. What Montgomery showed is that recruits with the II genotype had the biggest fitness increase after training, followed by the IDs and finally the DDs. In fact the DD group showed no improvement at all. Determining the *ACE* polymorphism is a simple procedure taking virtually no time and only requiring a saliva sample. Not surprisingly this work initiated a number of studies looking at athletic performances. Things are not all bad for the DDs. They seem to predominate in sports that require strength, rather than aerobic performance. So sprint swimmers are mostly Ds. The trend can be seen best when you compare the distances of running events; the longer the distance of the race the more likely you are to find a predominance of the 'I' polymorphism.[17]

Montgomery is an interesting character; an intensive care doctor and cardiovascular geneticist with a sideline in sports science. Unfortunately he has been hoist with his own petard. No elite mountaineer without at least one copy of the aerobic 'I' genotype has ever climbed above 8,000 m without the aid of oxygen. As luck would have it Montgomery, a keen mountaineer, is a DD, although he still holds out the hope of defying his genetic liability.

ACE is an acronym for *Angiotensin Converting Enzyme*. This enzyme converts one small protein called Angiotensin I to another called Angiotensin II. Angiotensin II triggers a constriction of blood vessels dropping blood flow and increasing blood pressure. The II polymorphism is associated with decreased *ACE* activity. Therefore it would be expected to prevent this drop in blood flow. Could this be why II people are better at long distance running and mountaineering—they can deliver more oxygen to their tissues because of an increased blood flow?

Nice though this theory is in principle, things are not quite so simple. For example there is a class of medical drugs that target *ACE* called *ACE* inhibitors. If you are having treatment for high blood pressure it is highly likely that you are on one of these drugs. An *ACE* inhibitor acts in a similar way to the II *ACE* polymorphism in that it reduces *ACE* activity. When taken by a patient it can drop blood pressure and increase blood flow. What about healthy individuals? Maybe an *ACE* inhibitor could increase blood flow and enhance long distance running performance, thus overcoming the shortcomings of the DD polymorphism? *ACE* inhibitors could be the key to getting Dr Montgomery to the top of Mount Everest. Unfortunately animal and patient studies point to the same negative conclusion. *ACE* inhibitors do not have performance effects; they do not magically turn DD performing people into II performing people.[18]

This observation highlights an important point. People who have a II polymorphism will have had it from womb to adulthood. This is very different to taking a drug that inhibits *ACE* activity in adult life. Living with the polymorphism the body will have responded during development. It is this combination of genes and the development environment that is the likely cause of the performance increase, possibly via subtle interactions with a host of other gene polymorphisms. It may not be possible to produce a similar effect by the brute force approach of targeting the *ACE* gene product in an adult human. It is possible that *ACE* gene doping in adult athletes will be no more effective than the pharmacology of *ACE* inhibition in enhancing athletic performance.

The initial excitement around *ACE* has subsided somewhat. More recent studies have not been quite as clear-cut with regards to performance. Interest has switched instead to a gene called *ACTN3*. This was the first gene polymorphism known to affect muscle function. At the gene level the effect is dramatic. The *ACTN3* gene codes for a protein called actinin-3. One polymorphism stops the formation of any of this protein so if you have inherited defunct genes from both your parents you cannot synthesise actinin-3 at all. The world is therefore divided into people who have no actinin-3, fifty per cent active actinin-3, and 100 per cent active actinin-3. In 2003 Nan Yang and his team at a Sydney hospital published a paper that showed that elite sprinters had a significantly higher frequency of the active actinin-3 phenotypes.[19] Indeed only eight per cent of the male elite sprinters were deficient. In a subsequent study of fifty-one athletes in power sports only a single individual had both copies of the dud gene.

What does actinin-3 do? The filaments that allow a muscle fibre to produce force are made up of two proteins—actin and myosin. Actinins are proteins that bind actin filaments to the Z lines which define each functional unit of muscle contraction—the sarcomere. For some reason that is not completely clear, the presence of actinin-3 favours the production of the fast twitch fibres, which are designed for high power output and which are more efficient at using energy using the metabolic pathways that don't need oxygen. Optimising these pathways is crucial for energy utilisation in power sports, hence the tendency for elite sprinters to have high levels of actinin-3. In contrast, slow twitch fibres are optimised for exercise in the presence of oxygen. Therefore all is not lost if you have no actinin-3. It is more likely you will be an endurance athlete.[20] Just like the *ACE* gene you can still be an elite athlete whatever your polymorphism—you just have to pick your sport. If you are *ACE* deficient and *ACTN3* sufficient you should try football or sprinting; if on the other hand you are *ACE* sufficient and *ACTN3* deficient, then long distance running or mountaineering could be the way to go.

So does this mean we can simply select sports teams on the basis of a quick genetic test? It used to be that people would just look at their children to suggest sports they should take up—you're tall what about trying basketball? Some countries took this to extremes with a sophisticated battery of tests of body shape and composition to try and optimise their potential for Olympic gold medals. However, whereas countries—and individuals—used to look at anatomy, gene screening is set to become the preferred tool for the twenty-first century.

There is evidence that this is already happening. And it is not only countries that are considering using genetic screening for talent identification. Society has historically had to cope with the ethics of testing for individuals with regard to genes that have a strong predisposition to a disease state. However, more and more non-life-threatening lifestyle choices are becoming influenced by a genetic measurement. In sport this is led by home testing for the *ACTN3* polymorphism, introduced at a cost of just £169 by Atlas Sports Genetic in 2008. Parents are being encouraged to choose the area of sporting endeavour for their children on the basis of this information.

The differences in the frequencies of gene polymorphisms are not just detectable when looking at elite athletes. They exist in the underlying populations as well. For example the prevalence of the 'strength promoting' *ACTN3* polymorphism varies in different countries. Australians have only a thirty per cent chance of having maximal actinin-3 activity, but in Jamaicans it is seventy per cent. Only one per cent of Jamaicans have zero levels of *ACTN3*.[21] This is interesting but it is hardly the 'smoking gun' for those who claim that the Jamaicans are the best sprinters. Usain Bolt blew away a field of black sprinters in the Olympic sprint finals. Not only did all his rivals in the finals have *ACTN3*, it is almost certain that all the white sprinters at the Olympics had the same advantage. So yes, it is indeed more likely that the average black Jamaican will have the basic genetic tools to become a sprinter when compared to a white Australian. But this cannot be the complete picture.

Just picking on a couple of genes—one that helps endurance sport and one that helps sprinters—is not enough to unravel the complexity of the genetics of sports performance. What about looking at all the optimised genes together? Alun Williams at Manchester Metropolitan University has started to look at this using statistical tools.[22] He took the top twenty-three known performance enhancing polymorphisms suggested to enhance endurance sport; he then combined this with their known prevalence in the population. A few simple(ish) sums later and you have the chance of finding someone with this magic combination. It seems very low (0.0005 per cent) or a 1 in 200,000 chance. However, there are a lot of people in the world. 1 in 200,000 of a seven billion population is still 35,000 people with 'perfect' genes. But there clearly aren't 35,000 super elite athletes that out-perform all others.

What is going on here? One obvious explanation is that we have not discovered all the required genes. Even adding a few more combinations could make the probability of finding the genetic superman very unlikely. Equally we don't know if there are critical interactions with the environment such that these 35,000 people have yet to be in the right place at the right time to compete at sport. Maybe the combination of these polymorphisms is linked to a gene for laziness so these superhumans are destined never to reach their full potential? This argument is no more fanciful than many of the ad hominem explanations I have heard in recent years.

What these studies actually tell us is that we have come full circle in our argument. We can now name the genes that have an effect on performance, but the environment is also critical. We can suggest beneficial combinations of genes, but these need to be matched with the right environment to optimise performance. In one sense we are just as ignorant as when we started out thinking about these matters, although hopefully we are ignorant in a more informed way.

My prejudice when I started writing this book was that any physical performance activity is going to be limited by a wide variety of factors. The over 200 performance enhancing gene polymorphisms chime with my prejudice. Changing any one is unlikely to have a dramatic effect as some other limitation will take over. Instead you probably need a range of simultaneous changes to even make a small difference in performance. Given the large human gene pool we have been selecting for sports performance the remaining performance enhancements are likely to be small and incremental.

This is analogous to my own field of biochemistry in the 1970s and 1980s. A new theory, called metabolic control theory, suggested a way of quantifying exactly which enzyme restricted which metabolic pathway.[23] If we could only alter this enzyme (maybe by altering the genes that made it) we could enhance the activity of the pathway. This idea was initially jumped on by the biotechnology industry as a godsend. Knowing the exact gene that was the bottleneck, they jumped in to change it, hoping to engineer their yeast and bacteria to make higher yields of useful biochemicals and drugs. In fact what happened was that as soon as one gene limitation was removed, another part of the pathway became a bottleneck; just as, after building a new lane of a motorway or freeway, congestion quickly moves to elsewhere in the system. Therefore a dramatic number of genetic mutations is necessary to get a small 'performance' increase; the law of diminishing returns soon set in. This seems to be the case for human athletic performance.

On the surface this is bad news for the gene dopers—and good news for the anti-doping authorities. We might be able to put genes for human growth hormone or human EPO in yeast or bacteria to trick them to making us supplies of drugs. But genetic engineering techniques are much harder to introduce in humans than they are in these simpler systems. Even with all the resources of medical science, only one gene is selected for modification at a time in human gene therapy. If athletic performance enhancement requires multiple gene changes, successful gene doping is a long way away.

Seventy years ago there was an argument about whether evolution proceeded by a small accumulation of mutations, subject to natural selection, or whether a single dramatic event could be responsible for the formation of new species. The term 'hopeful monster' was coined by Richard Goldschmidt in his 1940 book *The Material Basis of Evolution*.[24]

Darwinian evolutionists have largely ridiculed this view. But when it comes to gene doping a hopeful monster is exactly what is needed. Can a single gene change have a dramatic improvement on exercise performance? Or is gene doping doomed to failure from the start?

Humans with extraordinary genes

If at first glance the study of human polymorphisms seems to fly in the face of single genes being key to performance and provide little comfort for gene dopers, there are two notable exceptions. One relates to aerobic exercise and one to strength-based sports. The two proteins involved are those we have discussed previously—EPO and myostatin. A couple of extraordinary people with modifications in the genes coding for these proteins, have suggested that gene doping might not be so fanciful an idea after all.

EPO is the hormone that controls the production of red blood cells. At the end of the nineteenth century a Finnish man was born with a mutation in the receptor for EPO. This receptor is present in the bone marrow; when EPO binds it triggers the signal to make more red blood cells. But the body needs to know when to stop. The EPO receptor is turned off by a phosphorylation signal. This Finnish man had a gene mutation that made his receptor smaller. It retained the site for EPO binding that triggered the production of red blood cells, but it had lost the site where phosphorylation occurred. As a result the EPO signal could not be turned off. The result was an over-active EPO receptor and increased levels of red blood cells. The mutation was dominant and studies of his family tree showed that it passed successfully to the present day through five generations; over twenty-five affected individuals in the genealogy are known to have high levels of red blood cells (chapter 1 ref. 18).

Some mutations are good and some bad. Those that have good and bad characteristics frequently end up becoming fixed in the human population, like the sickle cell gene that protects against malaria.

Likewise having an excess of red blood cells could be good and bad. On the good side there would be an increase in aerobic fitness; but on the downside it could increase the risk of heart attacks and stroke as too many red blood cells can make the blood viscous and decrease flow. This is the reason why EPO doping is hazardous. However, at least in this Finnish family the downside seems manageable—there are no cases of heart problems in the five generations attributable to the increased number of red cells, perhaps due to compensatory changes occurring during development.

There is a good reason why this family tree has been so well studied. In 1937 its progenitor sired probably the most successful Olympic cross-country skier of all time—Eero Mantyranta. His genetic anomaly resulted in an excess of red blood cells which, combined with hard work and training of course, resulted in three Olympic and two World Championship gold medals, two Olympic and two World Championship silver medals and two Olympic and one World Championship bronze medals.

What about myostatin? Can a similar single mutation lead to similar riches for sprinters or weightlifters? In 1999 a German sprinter gave birth to a boy with abnormally sized muscles. The doctors were reminded of the overmuscled Belgian cattle that were deficient in myostatin. So they did a genetic test and, sure enough, a similar situation had arisen (chapter 5 ref. 6). The boy had inherited two copies of an altered myostatin gene that prevented formation of the protein. His mother had only one bad copy. Presumably the father had the same mutation but he was not made available for testing. The genetic history suggested feats of strength ran in the family; the boy's grandfather was able to unload kerbstones by hand for example. Although it is unclear whether his mother's sprinting career was helped by her reduced myostatin, it is clear that the boy is very strong. At the age of four he was able to hold two 3 kg dumbbells horizontally with his arms extended. Hopefully there will be no deleterious effects of his lack of myostatin. The studies from myostatin-deficient mice, cattle and dogs suggest he should live a normal lifespan. Whether he will be able to, or want to, make use of his strength in the field of sport is of course a different matter.

This German baby is not unique. Myostatin-related muscle hypertrophy is a recognised, if extremely rare, clinical condition. In the absence of general genetic testing we cannot be sure how prevalent this is among elite athletes. However, it is clear that babies with this condition can perform feats of surprising strength. The ultimate strength test for a gymnast using the rings apparatus is to hold his body vertically off the ground with his arms in a horizontal position. This 'iron cross' was accomplished readily by a five-month-old American myostatin-deficient baby.[25]

How this relates to elite performances in sports is another matter. We should be mindful of the Usain Bolt story. The world has woken up to how advantageous it can be for a sprinter to be tall as well as powerful. Following Bolt's example some people suggested that his would be the new shape of sprinting. However, it is unusual to have someone with Bolt's height and strength possess the motor control to coordinate the movement of their legs and arms at such high speeds. Athletics is a complex business and just possessing one alteration in anatomy may not be enough to guarantee success. Bolt's genetic combination of size and agility may be rarer than we think.

In a similar vein while someone with myostatin-related muscle hypertrophy may find it easier to put on muscle mass, they may have more difficulty making use of that in a competitive environment. For example mice that are deficient in myostatin may have larger muscles, but they produce a reduced force per unit area.[26] Quantity does not always come with quality.

Genetically engineered super rodents

In contrast to the more general studies of polymorphisms, the specific cases of the individuals with these dramatic myostatin and EPO modifications allows us to entertain the possibility that altering a single gene could dramatically improve performance. So is such a change possible in practical genetic engineering terms? A dramatic video accompanies a seminal paper published in 2007 by Richard Hanson at Case Western University.[27] It shows two mice running on a treadmill. The only difference between them is that one has an alteration to a single gene. The one with the added gene can run faster and longer. I was shocked when I saw this paper. Unlike my prediction it does indeed seem possible to create an animal that can perform much better in long distance running by just altering one gene, although it has to be admitted the gene alteration was not small. In fact the gene was made 100 times more effective and it was specifically targeted to muscle cells.

This genetic change was made in the DNA of the embryo and was therefore present throughout development; this can have dramatic knock-on consequences in unexpected areas. For example the 'supermouse' not only runs better, it is in general seven times more active in its cage and is much more aggressive. It eats significantly more food, remains fitter, has less fat, and lives longer. The mutant mouse can even have offspring when it is over two years old, double the age of most mice. It is highly unlikely you could get these diverse effects by modifying this gene in an adult animal. But in terms of sporting performance it is the biochemical consequences that are most dramatic. The key enhancement is that the supermouse is able to oxidise fats much more efficiently. This is the key to aerobic running performance. The result is that it can run without using up its carbohydrate stores; its muscle glycogen lasts longer and it produces less lactic acid. In marathon terms the supermouse never 'hits the wall'. It is not completely understood how this is achieved but it does have an increased content of the oxygen consuming mitochondria and a high concentration of a mobile fat store called trigylcerides.

You may be curious as to what gene can produce such diverse effects; I certainly was. It turns out to be a gene that codes for an enzyme called PEPCK-C. This came as a complete surprise as the enzyme was nowhere to be found on the list of possible performance enhancing genes. As far as we know no human has a polymorphism in PEPCK-C that improves performance. In fact the researchers who discovered the running effect were not looking for this at all; Hanson and his colleagues just noted the behavioural effects and decided they should follow where scientific serendipity had led.

Why did they choose this gene and why did they target the muscle? They say in their paper that 'Skeletal muscle was selected as a target organ because there is no clear indication of the metabolic outcome of having a high activity of PEPCK-C in this tissue'. In short they did the study pretty much just out of scientific curiosity, found a counterintuitive unusual result and followed it up to produce one of the seminal papers in performance research. Sometimes science just works that way, even though scientists are not always so keen to admit it.

PEPCK-C codes for an enzyme catalyst called phosphoenolpyruvate carboxykinase. When I was an undergraduate student I learnt that its main role was in the liver, where it aids in the conversion of lactic acid to glycogen—thus rebuilding the carbohydrate stores depleted during activities such as exercise. But like many things you learn when young this is not the whole picture. PEPCK-C also seems to be able to take a more active role in metabolism, increasing the concentration of intermediates that can act as a fuel supply for mitochondria. Somehow having the potential to have more active mitochondria during development has led to a fundamental switch in the behaviour of the mutant mice. There is probably some sort of positive feedback between the increased mitochondrial activity and the efficiency of that activity; the mice may have been reprogrammed to train longer and harder. But this is still speculation.

We should exercise caution when extrapolating these studies to humans. Unlike a modern day greyhound or racehorse, laboratory mice are purebred strains that are by no means optimised for long distance running. It is therefore relatively easy to selectively breed mice for enhanced running performance. Perhaps the optimal performance benefit we see in these mice may already be present in athletes like Paula Radcliffe that have arisen from the more diverse human gene pool.

Interesting as the PEPCK-C supermouse is, it is strictly relevant only to situations where parents might want to genetically alter their offspring. This happens occasionally, but in the direst clinical situations; for example genetic testing is done to discard embryos that carry defective genes that cause a life threatening disease. However, there are currently no plans—nor an ethical framework—to create positive mutations in test tube babies for enhanced health. It is likely that breeding mutant sportsmen is even more of a distant prospect.

However, what is not so distant is the idea of altering a current athlete to enhance their performance. This would not be a mutation that would be passed on to their peers; we are talking about targeting muscles not gonads. The research driver here has been medical. Researchers want to be able to enhance cell growth in patients who have abnormal development, such as those suffering from muscular dystrophy. But could we use the same principle to genetically engineer an increase in muscle mass in an athlete?

The pioneering work in this field has been done by Lee Sweeney at the University of Pennsylvania. He chose the gene called insulin growth factor (*IGF1*). This is one gene that turns on protein synthesis—the START signal. Unlike the STOP signal myostatin, which is only active in muscle cells, the IGF-1 protein works in all cells. Sweeney got round this problem by targeting his gene enhancement specifically to the muscles. He gene doped one leg of a young rat and used the other as the control.[28] He was able to show a fifteen per cent increase in muscle mass following the injections. The difference was even greater as the rats aged, suggesting that gene doping may be more effective in preventing age-induced muscle dysfunction, rather than enhancing optimum function when we are in the prime of life. Repeated injections were needed. This is good news and bad news for gene dopers. On the down side it is less convenient; injecting even once deep into the large muscles of an elite athlete is a major challenge compared to the small muscles of a rodent. On the other hand it does have the advantage that the gene injection is not permanent. If you get the timing right you could be 'clean' when you entered competition and knew you were going to be tested.

Hanson's super running mouse and Sweeney's mighty power rat put paid to what I thought was a sophisticated biochemical argument about the checks and balances in the control of human performance. Altering a single gene *can* have a dramatic effect on maximum performance in animal studies. This change does not even have to be in the embryo—a mature adult can be successfully doped. When it comes to gene doping it is the positive example of the superhumans with single modifications in EPO and myostatin modifications that holds the key, not the negative example of the complexity of interactions when studying genetic polymorphisms in populations.

These two mice are not isolated examples. In 2004 a US/Korean group were studying a gene called *PPAR-Delta*.[29] They found activating the gene in fat cells caused weight loss. Curious to look at the effect in muscle cells, they saw a strange result. *PPAR-Delta* caused a change in the fibre type, increasing the number of aerobic slow twitch fibres. This resulted in an animal that could run twice as long; they immediately dubbed it the 'marathon mouse'.

Can we move beyond rodents? One of Sweeney's colleagues at the University of Pennsylvania, Jim Wilson, injected the gene for EPO into the muscles of macaque monkeys. This, as expected, increased the red cell blood concentration. But there were two problems. At first the treatment was *too* successful. They had to regularly bleed the animals so that the blood did not become too thick to flow. But some animals had the opposite problem. Their bodies not only rejected the foreign EPO gene but in the process prevented the normal EPO gene from working. The result was a dramatic fall in EPO levels resulting in severe anaemia.[30] These studies show that of all the doping techniques we are talking about in this book, gene doping is currently by far the most technically difficult and risky to attempt. So how is it done and what can go wrong—or right?

How does gene doping work?

Gene doping in sport is a bastardised version of gene therapy, differing only in the desired outcome—in the one case curing disease, in the other improving athletic performance.[31] But the molecular tools used are the same. In this context it is interesting to note just how slow the progression of gene therapy has been. In the 1970s gene therapy was touted as heralding a new age of medicine. However, the technology has proved both difficult to implement and difficult to make safe. The simplest targets are disease caused by a single gene defect, where the addition of a 'normal' gene can restore healthy function—for example sickle cell disease, cystic fibrosis or muscular dystrophy. The new DNA is either added on its own or, more usually with the help of a viral vector. The vector can target the replacement DNA to the right place and allow it to merge with the host's DNA. As a normal role of a virus is to put its own DNA into the human chromosome, mixing the new DNA with a virus has proved reasonably efficient. Occasionally things go wrong though. The virus can insert itself into the host's DNA next to a cancer-inducing gene; this can cause leukaemia. Or the host's immune response can attack the foreign DNA; at best this renders the treatment ineffective, at worst it can cause the death of the patient. Having said that, these are extreme cases taken from over a thousand clinical trials.

As of 2011, over thirty-five years after it was first suggested, there is still no clinical condition for which gene therapy is a routine treatment. Lack of effectiveness is probably more of a long-term concern than the prevalence of adverse side-effects. The hope is that the efficacy of treatment can be improved and the side effects reduced. Thus the 1970s hype would at last be realised. The nagging possibility remains though that the treatments appear relatively safe precisely because the DNA does not insert into the host. That is their safety is linked to their ineffectiveness. This may limit gene therapy to only the most severe life-altering conditions. Still we know that many athletes are happy in principle to play fast-and-loose with their long-term health. In this context the risks in gene doping are probably no greater to those many athletes are already prepared to take. If you are prepared to experiment with blood thickening agents such as EPO or untested combinations of steroids and growth hormones, why should genetic injections hold any extra fear?

What are best genes to dope?

By this stage of the book you probably already have an idea yourself as to what genes you would target if put in the position of a doper. Not surprisingly top of the list are those genes that have proved successful in animal studies. Increasing the effectiveness of genes for EPO, PPAR delta or PEPCK could improve endurance, increasing IGF-1 could enhance strength. But this is only the tip of the genetic iceberg. Like house prices, gene numbers can go down as well as up. It is theoretically possible to knock out a limiting gene, rather than just up the effectiveness of a performance enhancing gene. This brings the STOP genes into play, muscle myostatin being the prime example.

Other candidates have not yet been tested in animals. Extra copies of the genes could be inserted that code for the enzymes involved in testosterone synthesis. Genes that control endorphin production could be enhanced to enable athletes to perform through the pain barrier. EPO increases the amount of oxygen in blood, but another protein, VEGF, controls new growth of the vascular system. A combination therapy could be envisioned where the number of blood vessels is increased with VEGF and the amount of oxygen each vessel carries enhanced with EPO.

The possibilities are, if not endless then at least enough to support the research budget of a small country. In fact the basic science is scarily cheap. Peter Schjerling from the Copenhagen Muscle Research Centre recently described how he used high school students to make a 'dummy' gene doping construct as part of a school project.[32] There is an analogy here with nuclear weapons. The basic theory of how to make one is widely understood. But the problem—and expense—lies in the fine details of the engineering feat involved. In gene doping the real expense is not in the initial creation of a doping agent, but the difficulty involved in trying to make a product that is effective and safe. And then this would normally be followed by the million pound clinical trials.

Making gene doping effective and safe for patient use is difficult. However, if you are only interested in something that works a little bit and is a little bit safe, then the experiments really do become possible in a garage laboratory. It is perhaps unfortunate in this context that EPO is used as a model system to test medical gene therapy methods. This is because it is easy to check if EPO has been activated in the body by looking at the number of red blood cells produced. So you can test whether your doping has worked. EPO gene doping studies are therefore becoming rapidly optimised; the toxicity issues first seen in the macaque monkeys have recently been overcome.[33]

It is perhaps ironic that that one of the best characterised and optimised systems of gene therapy is likely to be the very same one that gene dopers are desperate to get their hands on. Indeed there is evidence that this is already happening. A UK company called Oxford BioMedica developed an EPO gene therapy for anaemic patients with low red blood counts. The product—called Repoxygen—shot to fame in the 2006 trial of the German track coach Thomas Springstein when one of his emails was published stating, 'The new Repoxygen is hard to get. Please give me new instructions soon so that I can order the product before Christmas.'

It is unclear whether Springstein ever got his product. In fact Oxford BioMedica stopped developing the product at about the same time, perhaps causing the supply difficulties. Ironically, as a clinical treatment, Repoxygen has a built in safety valve that switches itself off when the red blood cell count gets to normal; it probably wouldn't have enabled supernormal red blood cell concentrations for an athlete anyway. Springstein was convicted of an offence he shouldn't have bothered committing. Still his email was one of the clearest indicators that coaches are serious about getting their hands on gene doping technology. One attraction is the generally held view that these methods are undetectable in doping tests. We shall find out later whether this is really the case.

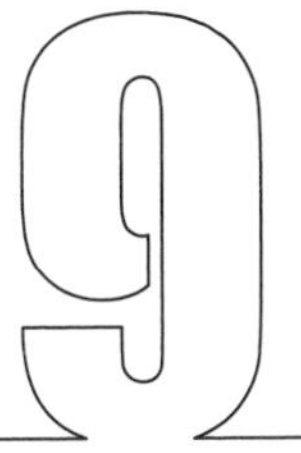

What Is Cheating?

'There is no good in arguing with the inevitable. The only argument available with an east wind is to put on your overcoat.'
James Russell Lowell

I have been a scientist for over 40 years—ever since I chose to dissect rather than eat the cod that my mother bought home from the fishmonger. But I am not a trained sociologist, philosopher or politician. So it would have been easy to write about doping control from a purely scientific perspective, focusing on how scientists try to measure drugs in an athlete's urine or blood and how athletes try to avoid detection. However, the story is not as simple as that. I want to explore the future direction of testing and the likely winners and losers in any war on performance-enhancing drugs. However, to do this—as in any war—it helps to know the objectives and the end game. For only if we know what we are fighting about can we know what tools we need to develop. Or indeed whether we should have a war at all. Perhaps as suggested by some it is elite sport itself that it the problem, and doping merely the symptom.[1] Maybe one cannot exist without the other?

Why wage a war on drugs in society?

As we have seen before the use of drugs in sport is intimately linked to their wider pharmacological and recreational use in society. So it is not surprising that many of the same arguments have been marshalled in defence of banning drugs in both sport and in the wider society.

Recreational drugs are illegal for a range of historical reasons influenced by politics, culture and religion. They can be divided into three main areas: the health of the individual concerned, the health of other individuals or the health of society at large; put crudely the personal, the political and the moral. The current political climate in debate in most secular Western countries revolves around health and crime; yet even then, it is underpinned by wider issues of morality.

The first reason for making recreational drugs illegal is paternalistic; it stresses the physical harm they can cause to an individual. The counterpoint relates to how appropriate it is to restrict liberty in the name of enhancing personal health. John Stuart Mill made the original case for individualism when he argued that an individual is sovereign over his body and that the only purpose over which power can be exercised over a member of a civilised community is to prevent harm to others.[2]

When it comes to the health of the individual, it might seem we have indeed chosen this path of non-interference. For example, paternalism has no place in extreme sporting activities. The arguments against mountain climbers or round-the-world solo yachters focus on the expense of a rescue if they get into trouble, rather than the dangerous activity itself. BASE jumping from buildings and other high places is one of the most dangerous extreme sports with a 1 in 60 annual chance of death. Yet the law does not focus on the health of the participant; the main legal threat aimed at those who jump from tall structures are charges of trespass or breaking and entering.

Yet when it comes to the population at large there are many cases of society acting paternalistically. Examples are not restricted to situations where it is deemed the person (e.g. a child) is not competent to judge what is in their own interests. Legislation is enacted frequently to protect an adult individual's health by denying access to a present pleasure on the grounds that it may ultimately prove harmful to them in the future. Perhaps surprisingly, given the regular criticism of the 'nanny state', people happily accept such laws forcing them to act healthily. So whilst legislating against mobile phone use in a car is clearly designed to reduce harm to other drivers, forcing people to wear seat belts is only of health benefit to the individuals concerned; likewise with motor cyclists and crash helmets. Yet failure to wear a seat belt in a car, or a crash helmet on a bike is illegal in many countries. Despite some complaints when these laws were first introduced, these restrictions on liberty are now commonly accepted and generally obeyed.

The second argument for criminalising the use of recreational drugs relates to whether the health of society as a whole is improved by making a drug illegal. Even the most extreme libertarians do not argue that we should allow people the freedom to harm others; the famous First amendment to the US constitution protecting free speech is not an excuse for someone to falsely shout 'FIRE!' in a crowded cinema.[3] So a legitimate case can be made that we should not permit a freedom (taking drugs) when that freedom results in quantifiable harm to others.

This debate tends to focus on more pragmatic grounds; for example whether making recreational drugs illegal would benefit the non-drug using section of society. Most of the harm recreational drug users do to others is quantified as increases in crime. Whether legalising drugs is a public good or evil therefore sometimes becomes a rather technical argument about police resources; is the cost of attempting to enforce the legislation justified by the reduction of crimes committed by drug users? Would legalisation increase or decrease anti-social activity and crime? In the UK reclassifying cannabis to make the penalties for its use less severe was justified to the public, at least in part, as enabling the police to focus their limited resources more efficiently on 'harder' drugs.[4]

The third case for maintaining a ban on recreational drug relates to the morality associated with their use. While altering the fine details of a criminal penalty is sometimes couched in rather technical statistics, this is rarely the case when arguments turn to complete decriminalisation or legalisation. As recreational drug use is viewed through a moral filter, discussions frequently become heated. This should come as little surprise. Drug use can alter the human emotional and spiritual, as well as physical state. It can change what we perceive, even for a short time and has the ability to affect who we are and how we view our place in the world.

In some religions the effects of drugs on mental state are viewed as a bad thing. Muslims are generally prohibited from taking alcohol in any form due to its potentially intoxicating qualities. The temperance movement, originating predominantly in nineteenth Century Protestantism, had similar goals. Yet whether it be the use of hemp vapour by the ancient Scythians[5] or the 1950s mescaline-induced visions of the humanist author Aldous Huxley,[6] drug taking has frequently been viewed as a profoundly enlightening spiritual experience. These practices have continued into the twenty-first century. For example members of the Native American Church in the USA are exempt from federal drug legislation relating to their use of the mescaline-containing cactus peyote.[7]

A pragmatic utilitarian view of drug legislation is clearly going to come up against these moral issues. Drugs have strong cultural resonance. This is not just a matter of organised religion; the hippies used drugs as part of a counter-culture movement. Indeed it can be argued that the use of LSD and cannabis in the 1960s as 'a mind detergent capable of washing away years of social programming,'[8] helped create much of the moral backlash that underpins the debate today in western society.

How does science inform legislation on recreational drugs?

Abolitionists often refer to the repeal of alcohol prohibition in the US in 1933 as how society could benefit from decriminalisation. Organised crime saw a decline in profits, with only a limited increase in alcohol intake by the population.[9] The response from the lawmakers is that alcohol use is not the same as heroin; the nature of the drug matters. Just like sport's anti-doping agencies, government lawmakers categorise drugs. In many countries the basic laws prohibiting drug use are old such as the USA's Comprehensive Drug Abuse and Prevention and Control Act of 1970 and the UK's Misuse of Drugs Act of 1971. What does change—again there are sporting parallels—is the list of drugs that fall under the act and the sanctions that are imposed for their (mis)use.

When it comes to recreational drugs politicians and the popular press seem to favour the moral argument. In the words of a 2007 report, UK law is 'driven more by "moral panic" than by a practical desire to reduce harm'.[10] This inevitably has the potential to lead to conflict when politicians, scientists and doctors interact. Nowhere was this more evident than in the case of David Nutt, who chaired the UK's Advisory Council on the Misuse of Drugs. Part of this council's remit is to advise government ministers on the scientific evidence that underpins the classification of drugs into three classes based on harmfulness. Possessing a Class A drug (the most harmful) is punishable by a maximum of seven years in prison; you can get up to five years for a Class B and two years for a Class C (the least harmful).

Nutt's most infamous defence of his views were in a paper[11] entitled 'Equasy—An overlooked addiction with implications for the current debate on drug harms'. It compared the harm from horse riding (equasy) to that from taking a current Class A drug (ecstasy). Nutt made the point that engaging in sporting activities with a horse is a far more dangerous activity that taking ecstasy; horse riding causes 1 adverse health consequence per 350 events compared to 1 per 10,000 with the drug. The point is clear even if Nutt is overzealous in places; counting the global warming effects caused by the methane emission from the horse as a harm to society is perhaps one of these instances.

What is less clear is what it really means to say that horse riding is more dangerous than ecstasy. For Nutt was well aware of the category error in a strict link between drug use and equestrianism. He was not—of course—calling for a ban on horse riding, but rather emphasising the sort of arguments that should be marshalled when discussing the 'harm' that a drug could do. However, it is clear that putting the argument in these terms is anathema to politicians. According to Nutt it led to the Home Secretary phoning him and accusing him of being a 'legaliser'.[12] The ensuing moral outrage coupled to similar proclamations from Nutt about the relative harm of cannabis compared to other drugs ultimately led to his removal from the Drugs Advisory Council he chaired. The feeling that this decision was informed by political, rather than scientific, judgement led to the resignation of five other scientists from the Council. The negative consequences for how UK governments access independent scientific advice are still being felt to this day.

I use this story as just one of many examples where governments, and by extension many people in society, find it difficult to act purely on scientific evidence of relative harm when it comes to drugs policy. So how do the arguments for banning drugs in sport compare to those society uses for banning recreational drugs?

Why wage a war on drugs in sport?

Barry Houlihan has summarised the historical arguments used to ban drugs in sport.[13] Firstly doping harms athletes; secondly doping is unfair to the athlete's competitors; thirdly doping undermines sport in society. Contrast this with the arguments against recreational drugs. Firstly drugs harm an individual; secondly drugs harm those with whom an individual interacts; thirdly drugs harm the moral structure of society. There are clear parallels in the first and third arguments. Even the second has resonance. For where a heroin user steals money directly from other people, a steroid doper steals gold medals and fame from other athletes.

Does sport have a unique nature that makes it easy to legislate without controversy? Sometimes the waters can seem just as muddy[13, 14]. For it doesn't need Nutt's 'equasy' example to point out the obvious;

athletes consciously and continuously put their health at risk without risking the ire of the sporting authorities. In some sports such as boxing the risks may be self-evident. However, they hold for any contact sport. In fact, it is a rare sport, contact or otherwise, that is not harmful to health at the elite level. Exercise may be good for you; elite sport demonstrably isn't.

Waddington and Smith illustrate this point with numerous relevant examples.[14] In the US, for example, the average length of career for an American football player is under four years, with injury being one of the main reasons for this lack of longevity. Between 1997 and 2007 on average five fatalities per year were directly attributable to American football, the majority of these being high school students.[15] In the round ball version, professional football players in the UK run a 1,000 times higher risk of injury than other so-called high-risk jobs such as construction and mining,[16] although it has to be said that in football incidents are only rarely life threatening. Trampolining at the highest level comes with an eighty per cent risk of stress incontinence. A UK House of Commons report in 2007 noted that, while it was difficult to ascertain the precise number of deaths caused by anabolic steroids worldwide, for it to be anywhere near the deaths caused by contact sports it would need to be in the hundreds or even thousands a year; it was suggested this was very unlikely to be the case.[17]

Not only is elite sport harmful to athletes, but it is also legal to take drugs that actively increase this harm. Most notably athletes regularly take high doses of anti-inflammatory drugs. Following retirement elite English football and cricket players like Gary Lineker[18] and Ian Botham[19] noted their relief at not having to deal with the stomach complaints associated with the side effects of these drugs. Botham claimed to have had to drink antacid solutions such as Gaviscon as if it were milk to counter the side effects of the drugs that enabled him to play in the first place. An even more extreme example is the case of Peter Elliott, the British athlete who won a silver medal in the 1988 Olympic games. He was only able to achieve this feat following five pain-killing injections in the space of seven days; he returned from the games on crutches.[13]

Basing a ban merely on drugs being harmful to athletes would of necessity result in elite sport itself being banned. So what about the second argument? Do drugs create an unfair playing field and so harm other non drug-taking athletes? Previous chapters in this book suggest that with the right genes and training regime drugs *may not* be necessary for top performance, at least in males. But this is clearly a debateable argument and certainly not true for female athletes. A sense of fairness is a powerful argument for fans who like to know that they are observing an event that is a true test of the athletes themselves. However, emotionally attractive as it is, basing a decision to ban drugs solely to level the playing field does not really stand up to close scrutiny. There are lots of cases where sport is not 'fair'. Athletes are always after 'the edge'—knowledge that will make the difference between success and failure. There is no desire to share this information with others. In most sports key information is a well-kept secret. Witness the British team that dominated the track cycling medals at the Beijing Olympics in 2008, claiming seven of the ten gold medals on offer. As well as a well-funded and well-oiled support infrastructure, this success was made more likely due to superior, though completely legal, equipment—termed 'technological doping' by its critics.

There is no doubt that being born in the UK in the 1980s meant that you were more likely to win a cycling gold medal in the Beijing Olympics than if you had the same genes, but were born in Morocco. Is it fair that rich athletes and countries can afford the best equipment? What about motor racing? It is well known in Formula 1 that the best driver doesn't always win—the car makes a difference. This contrasts with IndyCar racing in the USA where all the cars are of standard construction and driver skill is the key differential.

Even without such technological enhancement, professionalism has historically been considered akin to cheating. Until 2002, the Olympic Games was formally barred to athletes who were paid for their sporting performance. Sometimes even trying too hard went against the ethos of sport as a glorious amateur pursuit. Anyone who has seen the movie 'Chariots of Fire' will note the disdain that Harold Abrahams is treated with for hiring a professional coach, Sam Mussabini, to help him win the 100 m gold medal at the 1924 Olympics. Similar concerns about the ethics of pacemakers bedevilled Roger Bannister's attempts at breaking the four-minute mile.[20]

Technological doping gives you an inside line to success. Unlike with drugs there is nothing in the sporting rulebook that requires fairness of opportunity with regards to this side of an athlete's preparation. Take the example of a sports scientist who devises a training and nutritional regime that guarantees a winning advantage to an athlete; they then couple this to a new sort of running spike that enables their charge to run even faster. There is no obligation to make this information public. This advantage would not be obvious to the observer of the event who would be unaware how biased the race was. However, if that nutritional programme involves drugs then it is deemed unfair from the start.

Therefore the second argument—fairness—cannot be the sole reason for banning drugs. Otherwise sports would need to address a lot more than just the drugs issue to level the playing field. In some sports the playing field is levelled artificially. For example in the ancient Olympics there were no weight categories. But in the modern Olympiad many gold medals (e.g. boxing, weightlifting) are restricted to people of certain weights. And of course males and females are segregated. If fairness were the main reason drugs were banned then many sports would need to address the kinds of issues that boxing and weightlifting have already done with their weight categories. In fact making drugs freely available could be said to reduce unfairness in sport as everyone would be playing to the same rules.

What about the third argument? Does doping undermine the integrity of sport in society? Here we are in a complex ethical area. The use of drugs to enhance sports performance is a subset of the growing area of human enhancement.[21] This includes present realities such choosing the sex of your baby and future possibilities such as pills for IQ (Intelligence Quotient) enhancement or cybernetic implants. Ultimately this debate goes to the heart of what it is to be human. On the one hand we have the philosopher Michael Sandel who argues that enhancements reflect a drive to mastery and 'what the drive to mastery misses, and may even destroy is an appreciation of the gifted character of human powers and achievements'.[22] On the other hand we have futurists like Ray Kurzweil who view the eventual merger of human intelligence with technological implants as a 'slippery slope leading towards greater promise, not down into Nietzsche's abyss'.[23]

A similar debate exists in the sporting context. The Canadian Olympic medallist Angela Scheider has argued that doping undermines what it means to be human and therefore the strive for perfection in sport.[24] The alternative, more positive, view of sports doping has been advocated by Julian Savulescu who suggests that enhancing the chemical environment is no different than optimising any other aspects of the sporting process. Attempting to ban drugs does not only cheat athletes by creating an unfair playing field between those who choose to dope and those who don't, it cheats them by preventing them from realising their full potential.[25] As he says 'Performance enhancement is not against the spirit of sport; it is the spirit'. Drugs should be embraced rather than feared.

This is heady stuff. As a biochemist, I have some sympathy with Savulescu's views. Sometimes I am stressed and have trouble sleeping. So I take a pill to help me sleep, rather than dealing with the root cause of my stress. I may be too lazy to confront my problems, but I don't feel the pharmaceutical aid has made me any less human. Yet all drugs are singled out as being against the spirit of sport.

We should, however, pause before we slip too eagerly into the libertarian position. What would the sporting world look like in which there was no attempt to restrict chemical and genetic enhancements? In terms of male sport perhaps not too different, at least on the surface. But in terms of female sport widespread anabolic steroid use would affect both the spectacle of the sport and the nature of the people taking part. This causes disquiet. When questioned in a sports ethics class[26] an audience of students felt that the steroid use by Ben Johnson made him a 'cheat'. But female steroid users were labelled 'gender freaks'. Of course there are subliminal issues relating to sexism here—real women are not supposed to be overly muscular. Nevertheless the concern that androgenic steroids can change who you are as a person is clearly a live issue for many people. There will surely be similar areas of concern in the future. We may not like the sporting world we create by unleashing the full power of biochemistry and genomics.

How does sport control doping?

In the introductory chapter we saw the evolution of perceptions about performance enhancing drugs. Initially viewed as useful adjuncts to performance they subsequently became as demonised as a Class A drug such as heroin or cocaine. Historically the policing of drug use in sport was in the hands of individual sporting agencies, with the International Olympic Committee (IOC) at its fore. Many international sporting bodies are secretive and unaccountable. Occasionally light gets shone on the organisation by investigative journalists—witness the corruption that has occurred when bidding for host cities for the Olympics or football World Cup. Following Salt Lake City being chosen to host the 2002 Winter Olympics, allegations of bribery led to mass resignations and expulsions from both the organising committee and the IOC. Even as recently as 2010 members of the executive committee for football's governing body—FIFA—were suspended for breaches of the code of ethical conduct with regards to the choices of venues for the 2018 and 2022 World Cups.

While journalists have uncovered structural problems with the administration of high-level sport, it is perhaps significant that only when it comes to drugs do governments themselves get directly involved. In 1998 the French government via its customs and police force intervened in the Tour de France; many key team officials were charged with the criminal offence of supplying banned drugs to a sporting event. The IOC seemed to be losing control of the anti-doping battle to national agencies. In an attempt to assert their authority and set up their own worldwide anti-doping agency, the IOC convened a conference in their headquarters in Lausanne in 1999. In their book *An Introduction to Drugs in Sport* Ivan Waddington and Andy Smith (ref. 14) describe how the IOC were ambushed by the very same organisations and politicians they had invited to get approval for their new anti-doping agency. The UK Sports Minister, Tony Banks, led the criticism of the IOC; it was backed up by representatives from the USA, Canada, Australia, New Zealand and Norway. The European Union became involved, refusing to allow pharmaceutical companies and IOC sponsors to be part of any new body.

The result was the formation of the World Anti-Doping Agency (WADA) with a mission to 'promote, coordinate and monitor the fight against doping in sport in all its forms'. WADA effectively controls what constitutes doping—and how it is policed—in the vast majority of sports. The headquarters of WADA is sited in Montreal, well away from the IOC headquarters in Lausanne. The Chair of WADA alternates between governments, and the IOC and its funding, board and executive committee have 50:50 representation from the Olympic movement and governments. Given this governance structure it is not surprising that WADA's rules and regulations reflect governmental as well as sporting attitudes to drugs.

So what does WADA consider a doping offence and why? There are three rules governing WADA's anti-doping policy, mirroring the three arguments we have discussed earlier. Doping is considered to be:

1. the use of a substance or method that represents an actual or potential health risk to the athlete;
2. the use of a substance or method—alone or in combination with other substances or methods—that has the potential to enhance or enhances sport performance;
3. the use of a substance or method that violates the spirit of sport.

WADA define 'the spirit of sport' as the celebration of the human spirit, body and mind, characterised by values such as: ethics, fair play and honesty; health; excellence in performance; character and education; fun and joy; teamwork; dedication and commitment; respect for rules and laws; respect for self and other participants; courage; community and solidarity. We can only count our blessings that WADA is incorporated as a Swiss, rather than American, private law foundation; otherwise motherhood and apple pie would surely have been added to that list.

It is all too easy to mock. For we are back to where we started in Chapter 1 with the nineteenth century upper class Englishman and his concepts of fair play, equality and competition. However, the third rule is not just about individual integrity. Sport is a business. It needs to be marketed as an ethical competition so that people—and indeed sponsors—feel happy about being associated with it. In this context, rule 3 could be viewed as an exercise in pragmatism. It looks bad if all the athletes are perceived to be on cannabis and cocaine, ergo it is bad. Yet this is perhaps too cynical. Rule 3 is an attempt to put doping in sport in a wider moral context. Its presence forces us to think of the sort of sport we want. The absence of such a rule makes as clear a statement as its presence.

So we have three rules. Yet their implementation by WADA is somewhat surprising. Unlike baseball it is not a case of three strikes and you are out; unlike cricket one ball alone cannot lose your wicket. Breaking two rules is the magic number. So even if a drug can enhance performance it is not automatically banned. Caffeine for example is potentially performance enhancing, but is no risk to health and is assumed not to violate the spirit of sport. So it is fine to take, at least as long as it is the form of a beverage. Put in a pill, with its pharmaceutical overtones, and WADA start to get concerned (chapter 7 ref. 14)—the spirit of sport is being challenged.

Perhaps more surprising is the fact that a drug does not even have to be 'performance enhancing' to be on the banned list. It can get its two strikes by just being harmful to an individual and against the spirit of sport. WADA's interpretation of the latter has pretty much included any recreational drug that is illegal; hence the appearance of cannabis and heroin on the list, substances for which there is no evidence that they improve performance. Rule 3 yet again shows the link between drugs in sport and drugs in society. It seems that if a government bans something in society WADA will ban it in sport, whether or not it is performance enhancing.

While it is hard not to feel that this policy is influenced by the fifty per cent of WADA that is government run and financed, things are not that simple. For example at a House of Commons Science and Technology Select Committee in 2007 the British Minister of Sport, Richard Caborn, railed against WADA trying to police society as well as sport.[17] He called for WADA to look seriously at removing social drugs from their prohibited list as, in his view, WADA is there to 'root out cheats in sport'. The clear implication from the Minister is that it is the role of society—interpreted in his case as governments not WADA—to address drugs as a social issue.

As we have seen it is possible to provide counterpoints to each of the three rules underpinning banning drugs in sport. However, just because it is possible to pick holes in an argument, it does not mean that the opposite—permitting everything—is necessarily true. I agree with Houlihan here.[13] In Western society we hope our governments only make things illegal if they have a good reason to do so, rather than merely on a moral whim. In contrast, sport is full of arbitrary rules and we accept these all the time. It is here that the ethics of controlling doping in sport diverges from controlling drug use in society.

Some rules are obviously inherent to the sport itself. You are cheating if you pick the ball up and run in a football (soccer) match; you cannot throw a forward pass in a rugby match; you cannot run in a walking race. Some rules, though, are subtler. Baseball pitchers are not allowed to apply any fluid—whether spit or grease—on a baseball to impart changes in the trajectory to make it difficult for batters to hit. But it is perfectly legal, indeed a key part of the game, for cricket bowlers to spit and sweat on a ball in order to create conditions amenable for swing bowling, as long as they do not actively gouge bits out of the ball: similar sports, different rules. Sporting rules are not just arbitrary, they change with time; the spitball was not banned in baseball until 1920. Athletes break these rules relatively frequently. They are caught cheating in baseball (usually by using Vaseline) and in cricket (by gouging the ball with their fingernails or rubbing it roughly in the dirt). If they are caught they are punished appropriately.

Put in this context drugs could perhaps usefully be thought of as just another rule imposed by sport. As some sports—notably rugby—seem to alter their rules every year, the annual WADA banned list could be viewed as just that; a change in the rules. Of course I am not suggesting that rules against doping are of the same moral substance as rules that say football is a game of two halves, rather than four quarters. Nor that having quarters in American football makes it morally inferior to the game played in the rest of the world. However, in their essence sporting rules are pragmatic and part of the contract the athlete enters into with the sport. Can we take a similar pragmatic approach to drug taking? Does this make the ethics of drug use in sport less fraught than drug use in society?

Hence the pragmatic approach. The law is the law, ignorance is no defence, the authorities (in this case WADA) will punish you. Society will then judge what it feels about your punishment depending on how it views the crime. At the 2010 football World Cup in South Africa the Uruguayan player Luis Suarez illegally handled the ball to save a goal. This blatant piece of cheating prevented Ghana being the first African team to reach a World Cup semi final, making Suarez a villain in most of the world (apart from Uruguay of course). He broke the rules, was punished, and then society was able to make its opinion. The situation has clarity. However, this contrasts with another famous handball in the same World Cup year. The French star, Thierry Henry, handled the ball twice in the build up to a key French goal that knocked Ireland out of the cup. Though the offence was obvious to the TV cameras—and later admitted by Henry—the primary rule breaking was not detected by the referee at the time and was unpunished, leading to an unsatisfactory outcome. What the Thierry Henry incident tells us is that even a pragmatic, not overtly moralistic view of doping requires effective detection methods. Whether doping is viewed as evil or merely as a set of sporting rules to be obeyed, WADA must follow the same route. Create a list of banned substances and make sure you can get your scientists to detect anybody using them.

However, there are alternative proposals that would alter the whole rationale behind drug testing. Waddington[27] and Savulescu[25] argue that the sole purpose of anti-doping rules should be to protect the athletes' health. This idea that the only role a drug tester should play is in non-judgemental health checks has something in common with attempts to help drug users in society. Indeed, as noted by Waddington, in 1994 in Durham in the north of Britain sport and society's treatment of drugs coalesced.[27] A mobile needle replacement service designed to prevent HIV transmission in drug users attracted a preponderance (sixty per cent) of bodybuilders; the bodybuilders were after not only clean needles, but also advice about drug side effects. Subsequently DISCUS (Drugs in Sport Clinic and Users' Support) was set up specifically for this group of users. This example shows how legalising doping could be done in the context of reducing—or at least minimising—harm to athletes.

Yet would lifting a ban increase usage and therefore result in increased harm to athletes? This argument is frequently made with regards to recreational as well as performance enhancing drugs. For example it is argued that legalising cannabis would increase drug use. This is a live question in society. But in sport we know the answer; there are real statistics to support the fear that legalisation increases use. Drugs that come off the WADA banned list are put on a separate monitoring programme where they are still tested for, but no penalties accrue for a positive result. Both caffeine and pseudoephedrine levels increased dramatically in athletes following their removal from the banned list in 2004. This led to the return of pseudoephedrine to the banned list in 2010; it may yet do the same for caffeine.

There seems little doubt that removing all bans for doping would increase drug use in sport. It is also difficult to see that maintaining a paternalistic ban on doping in children, whilst freeing restrictions for adults would be practical. At the very least illicit use by children would increase in line with that in adults. It would also change the formal role of doctors and biochemists from guardians of drug-free sport to freelance advisers to sports teams, especially those that have funding to develop and test a drug programme. Academically I might relish the challenge; ethically I would fear it.

The question for those who feel that the health of the athlete is the key ethical issue is whether, as Waddington and Savulescu argue, bringing drug use into the open would be safer. Would more transparent medical monitoring compensate for increased use? This is a moot point. Athletes would still seek an edge. As they don't reveal their training schedules, it seems even less likely they would reveal their drug regimes. There is no guarantee that a doctor, scientist or coach willing to try out experimental combinations of drugs would be best placed to judge, or even care about, their long-term safety. Sporting bodies could not be seen to be actively managing a safe programme, if that programme involved the use of drugs that were illegal in the eyes of the government.

Pre-performance health checks, while of obvious benefit, would not stop potentially dangerous abuses in training. An individualised medical regime to safeguard all athletes outside competition would be even more expensive than the current random drug-testing paradigm. While allowing a biochemical free-for-all is ethically defensible—athletes are free individuals after all and can make informed choices about their health—the effects on the health of athletes are not trivial to predict. It seems likely that removing all restrictions on doping methods would on average do more harm than good; levelling the playing field to enable all to dope likely comes at some medical cost.

Whatever concerns we may have for the health of athletes the most common argument against doping remains that it is cheating. Yet why does doping have such disproportionate penalties, compared to other forms of cheating in sport? In football, diving in the penalty area when you have not been fouled can win you a penalty. Many players have admitted to trying to do this. If they are unlucky and get caught then they are given a yellow card. If they accumulate five yellow cards they are banned for one match. Contrast this to cheating by doping where you get a two-year ban for your first offence.

Why is doping so much worse? There are probably two reasons for this—one ethical and one more pragmatic. The ethical view is that many in society and sport have moral qualms about being associated with drugs, even without taking into account that it is bad for the image and bad for the sponsors. However a subtler, second argument can be made for these apparently harsh penalties. There is a tacit assumption that it is difficult to catch people for a doping offence. So when you do catch a cheat the punishment is extreme, in part to compensate for all the times they escaped capture in the past.

The role of the scientist

Scientists have an important role in sports doping. In the first place they advise what compounds should be on the WADA banned list. This is an analogous role to scientific committees that advise governments on the safety of recreational drugs. In sport though, the scientists must judge performance effects as well as health risks. They frequently err on the extreme side of caution. This means putting a strike against a drug if it has even the remotest chance of being performance-enhancing or detrimental to human health. This is probably wise advice in the context of WADA trying to keep the appearance of a drug-free sport.

Were a more liberal approach to be taken to drugs, the scientific role could become more challenging. For example Bostrom and Sandberg have proposed an 'Evolutionary Optimality Challenge' for any proposed performance enhancement, This asks the question 'why, if an enhancement is so good, have humans not evolved to possess it already?'[28] If there is no satisfactory answer then we should be cautious about proceeding. In this context the evolutionary adaptations to living at very high altitude are illuminating. The red blood cell content of native Andean Indians (Quechua) living in a village in Chile at 3,700 m was compared to those of Sherpas living in Tibet at the same altitude.[29]

The Quechua had evolved to combat the lack of oxygen in the air by increasing the number of red blood cells. This blood thickening is as high as is seen with EPO doping and is accompanied by similar potential problems with heart attack and stroke. In contrast the red blood cell increase was lower in the Tibetan Sherpas, who seem to have evolved to increase oxygen delivery by an alternative safer mechanism. So in this case an Evolutionary Optimality Challenge might suggest that it is inappropriate to use EPO as the means of enhancing blood oxygen delivery; there are clearly healthier biochemical and physiological ways to achieve this same goal. However, intriguing as such an approach might be to academic biochemists, it is highly unlikely that such convenient examples will exist for most performance-enhancing drugs.

Scientists have a second role in the doping process that is more high profile—developing the methods and implementing the doping tests that result in athletes being stripped of their medals and banned from competition. As we shall see in the following chapter this role comes with its own significant technical and ethical challenges.

10 Catching The Cheats

'The quickest way of ending a war is to lose it'
George Orwell

Scientists analyse our bodily fluids for all sorts of reasons. We have blood tests to check iron levels, plasma tests for cholesterol, urine tests for kidney function and faecal tests for bacterial infection. Do-it-yourself home testing kits—originally limited to pregnancy—are now a big money market, catering to the worried well. Increasingly employers are investigating urine samples from current and prospective staff for the presence of recreational drugs. In much of the world this procedure is restricted to industries where drug abuse can endanger safety such as transport or construction. However, in the USA most major companies and government departments drug test prospective employees.

But when it comes to workplace testing, even the US government is a neophyte compared to WADA. This is one area where the sophistication and funding is greater in sports than elsewhere in society. The earliest tests—both for sports and society—were based on our body's own detection tools for molecular invaders, the antibodies in our immune system. Proteins can bind very tightly to other proteins as we saw when looking at signalling cascades. And one of the tightest of these interactions is when an antibody in the immune system binds to a foreign molecule. As long as a drug is viewed as foreign, the body will mount a defence to attack it and remove it from its system. So it is possible to inject a drug into an animal such as a rabbit and harvest from its blood antibodies that bind specifically to that drug. The antibody is then used as part of a detection system.

Although the details of these methods vary they all have at their heart what is called a competition assay. A modified form of the drug is created in the laboratory that gives off a unique signal only when it is bound to the antibody. Into this mixture is added the urine of the suspect. If the urine contains a drug this will compete with the modified drug for binding to the antibody. The normal drug in the suspect urine is not modified to give off the signal. Therefore the more drug that is in the urine, the more the signal decreases, This is easier to visualise using a diagram (see Figure 24).

The fine details of measuring the signal have been modified over time. Originally a radioactive version of the drug would be made and the amount of radioactivity left after incubating with the competing antibody would be measured. However, biologists like to replace radioactivity in their assays as much as possible, partly for convenience, partly for reducing costs and partly for decreasing the amount of health and safety legislation to implement.

When I arrived in King's College London for my first scientific research job in 1989, I started working on a machine called an Electron Spin Resonance (ESR) spectrometer made by a US company called Varian. I was using it to look for reactive free radical molecules in the body for the purposes of medical research. But the technique could also be used to measure how fast synthetic chemical radicals tumbled in solution. This tumbling rate—and hence the signal size—is modified by binding to a much larger molecule, such as an antibody. I found out that in the 1970s the main purchaser of these machines was the US army. They synthesised a form of heroin that had been modified by binding to a free radical. How fast this radical tumbled depended on whether it was bound to an antibody or not. If a person had heroin in their urine then some of this would bind to the antibody instead of the synthetic version. Hence an ESR machine could detect how much of drug was present using a competition assay.[1] As opposed to the radioactivity measurement, the ESR result could be done very quickly and soldiers could be screened prior to leaving Vietnam. With money no object, the US Department of Defence shipped a batch of these heavy (and expensive) machines to Asia to ensure that no soldier left for home unless they had first been 'detoxified'.

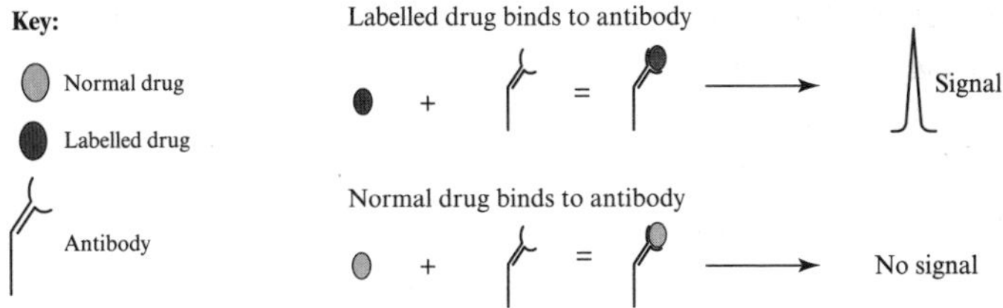

The laboratory mixes a labelled drug with the athlete's urine. Then it adds the mixture to the antibody. If the urine contains the drug it competes for the antibody. Less labelled drug binds and the signal goes down.

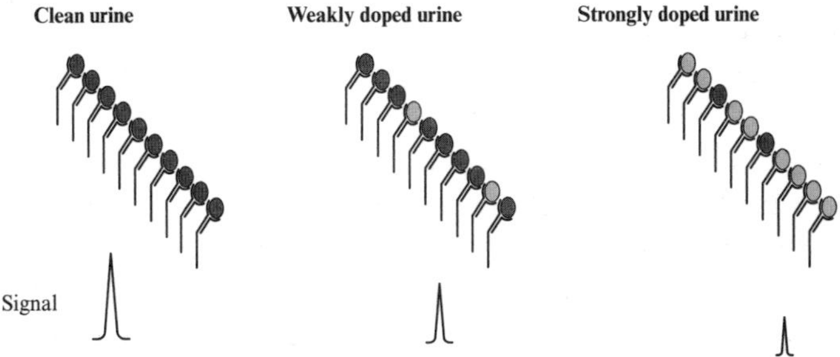

Figure 24 A competition assay for detecting drugs

A drug with a label is used as a test standard. If the athlete's sample (urine or blood) contains any of the drug it will compete with the labelled drug and reduce the signal.

Antibodies were used for the first testing programmes in athletes at the 1966 football World Cup in England. The winter and summer Olympics followed suit in 1968, though without conspicuous success—one athlete tested positive and that was only for alcohol. While these biological assays work and can be very specific, they do have some problems. When they were being developed in the 1960s and early 1970s, antibody production required live animals. Each time a batch of antibodies was made they would have slightly different properties and the assay would need to be recalibrated. By the 1980s this would become unnecessary, as identikit monoclonal antibodies became available from cells grown in the laboratory.

Even using monoclonal antibodies, biological-based tests require a separate antibody and a separate time-consuming assay for each molecule of interest. A new method was needed that was general enough to test for many drugs and quick enough to test many athletes. By the early 1980s this had happened and by the 1984 Olympics it could be reasonably claimed that a wide range of banned substances—now including anabolic steroids—could be tested for in the laboratory. The method used was chromatography or 'colour writing'. The principle of all chromatographic methods is that a mixture of molecules is passed through a material that impedes their progress. However, different molecules are impeded differently. Therefore the moving mixture (called the mobile phase) splits into its constituent elements as it passes through the material (called the stationary phase). All that is needed then is an assay for the different molecules once they have been separated. This is much easier than trying to measure them in the mixture. Again a diagram best explains this (see Figure 25).

You have probably witnessed a chromatography experiment. Paper chromatography is a favourite demonstration for trips to science departments during school open days. Ink is spotted on some filter paper, which is dipped in a solution of water. The ink dissolves in the water and moves up the paper. However, ink is a complex mixture of dyes that move at different speeds through the stationary phase (the paper). Soon different colours can be seen at different places on the paper. While paper chromatography is not widely used today in analytical science, it is of historical importance. For example at the turn of the last century it was used to separate plant pigments and determine the chemical nature of the green pigment chlorophyll.

However, chromatography goes far beyond this. It is the basis of over ninety per cent of all current measurements in chemistry or biology and is the basis of most of the drug tests used in today's laboratories. Even though the basic principles are the same, the nature of the technology has moved on. Scientists do not look at coloured spots on filter paper. Paper chromatography has been replaced by gas and liquid chromatography. In gas chromatography (GC) the mobile phase is an inert gas such as helium; in liquid phase (LC) it is usually an organic solvent. The number of molecules that can be measured in one sample of an athlete's urine is limited only by how good the separation can be made. In principle thousands of molecules can be tested in one run of a machine.

We should not become too carried away with enthusiasm. The laboratories of forensic science TV shows are filled with expensive chromatography equipment that provide answers in the time it takes for an advertising break. However, don't think that it is possible to put an unknown mixture of molecules into a machine and five minutes later know what each molecule is and where it came from—nor that this instant assay would stand up as evidence in a court of law. Thousands of peaks appear in a chromatogram. There is no way of knowing which one could be a banned steroid or stimulant. The only way around this is to have a standard sample of the drug. You then inject this into the same machine under exactly the same conditions. Only then can you know where to look for a peak in your urine chromatogram. So in order to test for a new drug you must have a sample of that drug in the first place. This is why the designer steroid THG was undetectable until the coach Trevor Graham anonymously mailed a few drops in a syringe to the Colorado home of Rich Wanninger, the Director of Communication at the US Anti-Doping Agency. Wanninger then passed the THG on to Don Catlin at the Los Angeles anti-doping lab. This one syringe led to

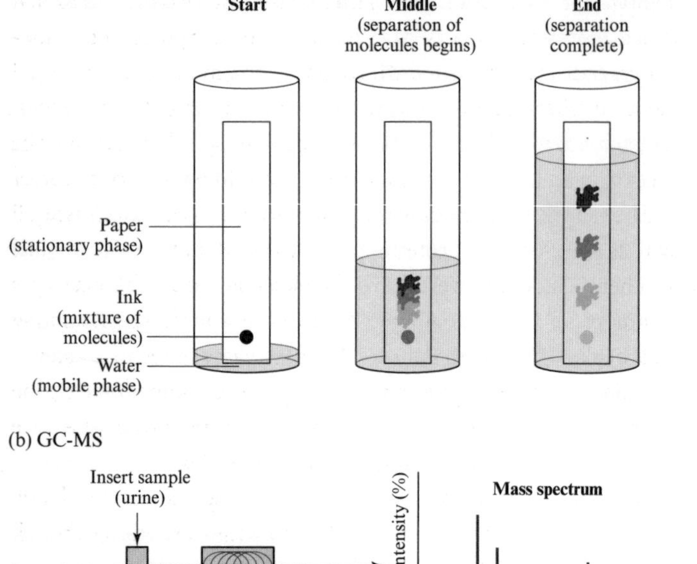

Figure 25 Using chromatography to separate compounds

(a). Shows how paper chromatography is used to separate dye colours in an ink solution. Detection is via optics (the eye).

(b). Shows how gas chromatography is coupled to mass spectrometry (GC-MS) to separate and detect drugs in an athlete's urine. Detection is via a mass spectrometer (an expensive machine).

the BALCO drug scandal that would eventually engulf the careers of track athletes Marion Jones, Tim Montgomery, Dwain Chambers, and baseball star Barry Bonds.

There is one further requirement to develop a robust and reliable test. The first GC methods in chemistry used a general detector that measured any molecule with a carbon atom in it. This of course includes all biological molecules as life is carbon based. So two molecules that that had the same interaction with the stationary phase and consequently moved at the same speed through the column would be detected as identical. There is always a worry that an athlete might have an unusual metabolite in their urine that ran at the same speed as the drug. The chance of measuring the wrong molecule and ruining someone's career would be too great if a general carbon detector was used. That is why all drug tests use a 'belt and braces' approach. The GC or the LC is coupled to another method for investigating the molecule. Again the analogy is with our paper chromatography. We can see a molecule separating from the ink mixture with our eyes. It looks yellow or red or whatever. We would not notice a compound that reached the same place on the paper that had no colour. And if we saw something green where we expected a blue colour we would suspect a contaminant.

Although the human eye is a powerful spectroscopic analytical tool, most drugs are colourless. However, they all weigh something, they all have mass. Therefore the output of the GC or LC can be coupled to a mass spectrometer (MS). The MS calculates the mass of all the peaks that have been detected. The position of the peak, coupled with the mass measurement, guarantees the identity of the molecule. In doping control—as indeed in forensic science—the coupled techniques of GC-MS and LC-MS have become the standard method of identification of small molecules.

The cutting edge: doper versus tester

GC-MS and LC-MS are great for detecting foreign compounds; their only limitation being the availability of standard test samples. However, it is difficult to detect a doping agent that is a normal body metabolite. Take the example of an athlete doping with actual human testosterone rather than an artificial steroid. The product in the urine is the same. The difficulty comes in trying to prove that the source is artificial.

One possibility is to set an arbitrary limit; above this it is assumed an athlete is doping. But there are problems in setting these upper limits given the wide range of natural variation in a population. This is especially a problem with testosterone where not only is there variability in natural levels in the body, but also urine production can vary by as much as 100-fold between individuals. Clearly setting a limit on the amount of testosterone in the urine is impossible. For one thing an athlete whose only crime was having a faster rate of transferring testosterone to his urine might end up failing such a test.

Instead scientists use the trick of measuring the ratio of testosterone to epitestosterone. Epitestosterone (TE) is a natural version of testosterone (T) but differs in that it is unable to enhance muscle development. In normal metabolism we make equal amounts of T and TE—the T:TE ratio in a drug test is close to one. But if you inject testosterone (T) you will increase the T:TE ratio. So it is this ratio that is measured in doping control, rather than the absolute testosterone level. If T:TE is greater than four the athlete is submitted to further tests that can result in a ban.

Not all natural compounds have such a good internal standard though. This is especially true for large peptides and proteins; these have the added complication that they are harder to detect using GC-MS and LC-MS methods. Yet measuring proteins and peptides is a key challenge, as this class of molecules includes some of the most relevant doping agents such as erythropoietin (EPO) and human growth hormone (HGH).

Artificial means of manufacture have made EPO and HGH widely available. But the artificiality contains the key to their detection. Synthetic peptides and proteins are made in large numbers by genetically modified bacteria. The genetically modified (GM) recombinant proteins produced in this way are subtly different from the natural ones. In the case of EPO this is because sugar groups attach themselves to the protein in the cell. The exact details of these groups depend on the nature of the cell that makes the EPO. Although the protein backbone is identical, natural EPO made in a kidney cell has a different pattern of sugars attached to it than recombinant EPO (rEPO) made in bacteria.

This difference is the basis for separation of natural and synthetic EPO forms by a technique called isoelectric focusing. The starting point for this test is to take a urine sample and suspend it in a gel between a positive and negative charge. If the protein of interest is negatively charged in the first place, it will start moving towards the electrode that has the opposite charge (positive). The trick is that the gel it is moving through varies in acidity; it is designed to be more acid the nearer the protein is to the positive electrode. Acidity means an increase in protons. Protons are positively charged. As they bind to the negative charges on the protein these become neutralised. Eventually over the course of several hours the protein comes to a halt.

The protein stops moving at a point when it is no longer attracted to either the positive or negative electrodes; its charge is zero (hence the term isolectric). The gel now contains a row of proteins that are frozen in space (hence the term focused). The position of the proteins depends on their initial overall charge (see Figure 26). As the sugars stuck on the EPO molecule contribute to this charge, different patterns of sugar binding give rise to different profiles on the final gel. However, a specific test for the EPO molecule is still needed; many other proteins in the urine might have the same charge. To enable this—belt and braces again—a monoclonal antibody specific to EPO is attached to the gel so that only proteins that can bind to this antibody are visualized.[2]

Some proteins are present in such small amounts that they can't be assayed in urine. For example, despite years of trying and over a million dollars of investment biotechnology companies and academic research scientists have failed to develop a test for human growth hormone in urine. However, the HGH concentration is over 1,000 times greater in blood than in urine, leading to the development of a successful blood test by a German endocrinologist Christian Strasburger.[3] As is the case with testosterone, this takes advantage of the fact that the body makes multiple forms of growth hormone. As only one of these forms is present in the recombinant drug, a simple ratio test can reveal whether a significant fraction of the HGH in the body is artificial. One antibody binds to the same form as the recombinant protein; another binds to all forms of the growth hormone. An increased ratio results in a failed test. Despite a lot of controversy as to its sensitivity in 2010 this test claimed its first positive results; a UK rugby league player and a Canadian student footballer.

Being able to test blood, rather than just urine, dramatically increases the tools an anti-doping scientists has in their armoury. Sampling blood doesn't just extend the reach of doping tests to

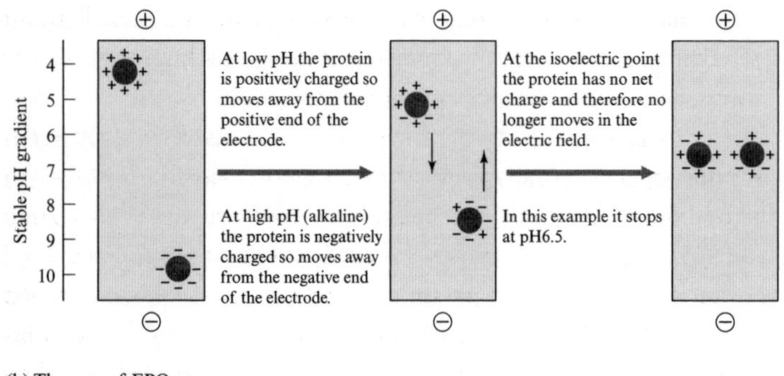

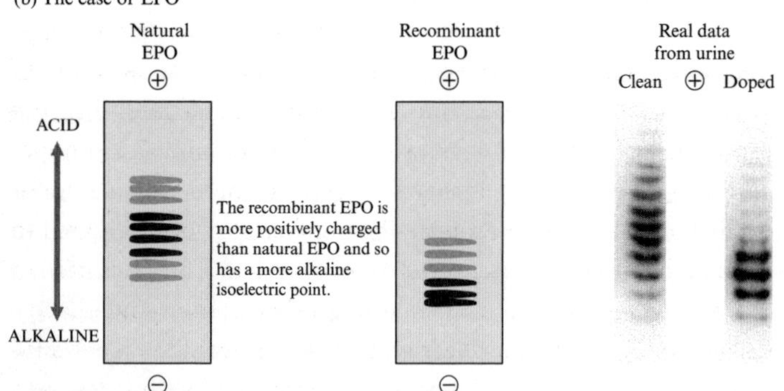

Figure 26 Detecting EPO

(a). The principle of using isoelectric focusing to detect a protein hormone.

(b). The complexity of using isoelectric focusing to detect EPO.

molecules that are not present at high enough concentrations in urine; it also enables completely different types of tests. Not surprisingly one of these is a test for blood doping. This involves many of the same methods used when a blood sample is sent to a medical haematology laboratory. The assay takes advantage of the fact that the surface of a red blood cell of each person contains different molecules. These are the famous blood groups of which two—the ABO and rhesus systems—are matched in a blood transfusion. However, it is less well known that over 300 different blood groups have been identified. These minor blood group constituents are not present in high enough concentrations to need matching for a blood transfusion. But they ensure that a full blood group profile is specific to a very small group of individuals. In the days before DNA testing, matching these minor groups was used to test for paternity in lawsuits.

In a modern laboratory the tester can make use of these very specific minor blood groups to determine if there is any foreign blood in the body. The technique is called fluorescence activated cell sorting (FACS).[4] It makes use of a range of antibodies that are specific to all the blood group types, major and minor. Molecules are then attached to these antibodies that give off different colours of light when illuminated—a process called fluorescence. The red blood cells can then be sorted by light and the number containing different blood groups detected. If foreign blood is present there will be a subpopulation of red cells that will be sorted into a different compartment because these doped cells have a different set of blood groups. Can a doper fool this blood test? Only if they have received blood from a donor matched to all the blood groups. The number of people required to screen to find such a perfect match would be significant. It is hard to foresee it being attempted by anything other than a government-managed programme, the sheer scale of which would make it difficult to conceal. Therefore the only way to reliably fool this test is to make use of identical (autologous) blood. For those athletes not fortunate to have a twin, this requires storing their own blood for later reinfusion.

Developing new doping tests—whether using blood or urine—means overcoming many challenges. One current problem, for example, is the difficulty in detecting drugs that only last a very short time in the body. Still most scientists would agree that given enough time and money new tests of the type described above could be developed as required when new doping agents came on the market. However, what really scares them is gene doping. This is likely to require the development of completely new methodologies.

Yet real progress is being made to detect gene doping. Take the EPO product developed for gene therapy—Repoxygen. This consists of a different, truncated version of the gene. A simple gene sequence test could reveal the source. Gene sequencing is a relatively simple form of analytical chemistry as it only seeks to measure a linear 2-dimensional code. The drive by scientists and companies to reduce the cost of DNA sequencing means that in five to ten years time (possibly sooner) it will be possible to sequence the entire genome of an athlete cheaply and conveniently. For any commercially available gene modification, manufacturers are required to provide the regulatory authorities sequence information of their product, along with details of how the DNA was inserted into the vector used to target the host's DNA. If this, information were made readily accessible to other agencies and scientists, new gene therapies should be detectable.

Assuming the commercial and regulatory issues are overcome, there is still one remaining problem. How do we obtain the DNA to test? In the sporting arena gene doping is likely to target specific tissue types in adults, rather than all cells in the developing embryo. If the gene is targeted to muscle—as surely it will be for genes aimed at improving strength—it may be difficult to detect without a muscle biopsy. Some sports still have difficulties in persuading athletes to donate for a blood test; muscle biopsies are very unlikely ever to be sanctioned.

All is not lost, however. There is a possibility that doping could be detectable in a blood sample even if the gene has been targeted to another organ. In 2004 the same French team, led by Françoise Lasne, that developed the original urine test for EPO explored this idea. They showed that injecting the EPO gene into muscle produced a form of the protein that differed from that made naturally in the kidney; this difference could then be detected in the blood using their normal test for the EPO protein.

Then in 2011 German scientists went further and showed that it was possible to create a direct blood test for gene doping.[5] Even if a gene is targeted to a different organ, a tiny amount of the recombinant DNA ends up in the blood. The trick is then detecting that this DNA is indeed the alien form that has been injected, rather than that from the athlete's normal cells. In a natural gene the DNA that codes for a protein (called an exon) is broken up with parts of DNA that are essentially non functional (introns). The gene is then spliced back together to make the protein. However, a synthetic recombinant gene lacks these non-coding introns. By creating a probe that binds a length of DNA spanning two exons, the normal DNA will not be targeted as it is protected by its intron. The recombinant DNA can then be amplified and detected (see Figure 27).

The research showed that many candidate genes discussed in Chapter 8 could be detected in this manner, including VEGF, erythropoietin, growth hormone, insulin-like growth factor 1 and follistatin. There is of course a counter strategy to this—doping with a gene made like the natural one, complete with introns. But while this is theoretically possible, it would require a significant investment of time and money from the dopers. No more could they rely on stealing ideas from the pharmaceutical or biotechnology industry, for there is no desire to create undetectable genes in normal medical therapy.

Trying to get away with it—or not

Some sports claim they are built around honour; testing is just not an issue. In 2005 Britain's most successful golfer Nick Faldo was happy to state[6] that 'Golf's been clean forever, probably because we've proven there's nothing out there we can take to enhance our performance'.

Even if there were such a substance golfers would never cheat as 'if you want to play golf, you forget about cheating from day one.' These words were echoed by golf's top administrators. PGA Tour commissioner Tim Finchem stated that even if research found there were performance-enhancing drugs it would be up to the players not to use them. In his words, 'The way you run golf is to pass a rule, and then you expect everyone to adhere to the rule. If we had reason to believe there was a violation, then we could resort to testing.'

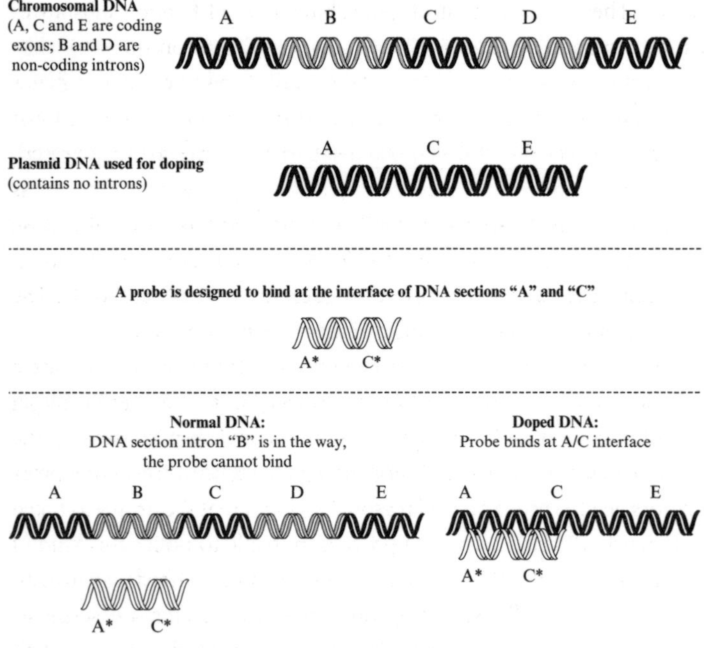

Figure 27 Detecting gene doping

A normal gene has introns that need to be spliced out of it. A doped gene has no introns. A probe can therefore be designed to detect the abnormal gene. Amplification of the signal means even the tiny amount of DNA that ends up in the blood can be detected.

Golf may be that rarest of sports where people never cheat. However, that does not mean there have not been accusations of doping, the most high profile being from the former player, and nine time major winner, Gary Player.[7] In 2008 golf gave in and finally introduced drug testing. Whatever the case with golf, it is clear that in several other sports not only is cheating endemic, but that the cheats definitely have no honour. They never admit they take drugs until the evidence is incontrovertible (and sometimes not even then) and they go to extraordinary lengths to avoid being caught.

This has resulted in WADA having perhaps the most draconian enforcement rules of any private organisation, or indeed many governments. These rules consist of: publishing a list of banned substances each year; banning retrospectively compounds not on the list if they can perform similar functions; and demanding to know where athletes are at any time so they can be tested. Most crucially an athlete must assume a 'strict liability' for the chemicals in his body. If a drug is present you are guilty; there is no need to prove how it got there. The strictness of these rules is the source of conflict in many sports, especially those like football, baseball and American football that do not need the Olympic imprimatur and so can afford to question WADA's authority. The implementation of novel testing can therefore be very slow.[8]

So why have such strict rules been imposed? Put simply they are a function of drug testing. In Western society possession of an illegal drug in your house is enough to secure a conviction. Possessing the drug in your body, however, is not enough; to secure a conviction prosecutors need to prove beyond reasonable doubt that the drug was taken voluntarily. However, it would prove impossible to police this kind of regime in sport. WADA does not have—nor could it afford—an investigative team of enforcement agents. Instead its agents are scientists and the policy of strict liability is the only weapon they have at their disposal.

This policing cannot be restricted to competitions. Many drug tests need to be administered within a few days of the drug being taken. Yet an athlete can stop taking a drug many days or weeks before an event and still retain a performance benefit; athletes could therefore gain the benefits of the drugs while ensuring they tested negative for championships. In the case of East Germany, the government even went so far as to drug test their athletes themselves before they left for competition to ensure there were no untoward surprises at the event. Tests for anabolic steroids only really became meaningful when random out of competition tests were introduced. With EPO the situation was even more extreme as the drug is only detectable in the urine for at most a few days after use. Yet the positive effects are long lasting.

Hence the controversial 'whereabouts' rule where athletes have to notify testing agencies of their location at all times. Missing three tests is equivalent to being convicted of doping and a ban ensues. Walking away from the testers is equivalent to being guilty. Here there is a parallel to the world outside sport. If you refuse to provide a sample for an alcohol test when the police have suspicion that you are drunk driving, the penalty is as severe as if you failed the test. There is an important difference though. The police do not take breath tests of everyone driving down the road. In contrast WADA take random drug tests of all athletes; in sport you have to prove your innocence. In recent years the English football captain, Rio Ferdinand and the World and Olympic 400 m athletics champion, Christine Ohoroghou, have both fallen foul of these rules and received bans despite not failing any drug tests.

The WADA code requires elite athletes to give notice of their location on a chosen one-hour period each day, seven days a week, both in season and out of season. A clue as to the reason for this strictness comes in a letter sent by BALCO owner Victor Conte to the UK sprinter Dwain Chambers about how athletes use the 'duck and dive' technique (chapter 1 ref. 9):

> First, the athlete repeatedly calls their own cell phone until the message capacity is full. This way the athlete can claim to the testers that they didn't get a message when they finally decide to make themselves available. Secondly, they provide incorrect information on their whereabouts form. They say they are

> going to one place and then go to another. Thereafter, they start using testosterone, growth hormone and other drugs for a short cycle of two to three weeks. After the athlete discontinues using the drugs for a few days and they know that they will test clean, they become available and resume training at their regular facility.

As it needs three missed tests to trigger a ban, after the second avoidance they have to stop taking steroids. However, it rarely comes to that if they play their cards—or rather mobile phones—right. These sort of techniques are the reason why, even with the more aggressive testing regime currently implemented, so many athletes never test positive for drugs despite a career of (mis)use. It has been estimated by sports scientists such as Perikles Simon at the University of Mainz that it would take as many as 150 tests to catch the average drug user.[9] It is worth noting that the (in)famous sprinters Marion Jones and Tim Montgomery never failed drug tests, and were only forced to confess their steroid use as part of criminal investigations.

It is not only in the timing of the test that WADA have to be vigilant. As well as having to reveal where they are on holiday, athletes also have their privacy intruded upon during more intimate events. When a urine sample is taken there must be a chaperone present to 'act as a witness of sample provision'. This is necessary because athletes have a long history of trying to pass off other people's 'clean' urine samples as their own. Perhaps not surprisingly cyclists at the Tour de France pioneered some of the more interesting avoidance techniques (chapter 1 ref. 2). Initially the attempts were primitive. Cyclists kept clean urine in a plastic bulb under their arm with a rubber tube running down inside their sleeve. When a sample was needed they just unstoppered the cork at the end. This worked fine until there was a blockage in the tube; the resultant physical inspection led to the disqualification of the 1978 Tour de France leader, Michel Pollentier. (Pollentier later claimed that he only tried to use someone else's urine as he was unable to supply his own).

Wary of being caught by eagle-eyed inspectors, cyclists turned instead to keeping the tube inside a condom hidden in their anus. The part of the tube facing out would be covered with carpet hair to fool an over eager inspector. An added bonus was that the drug-free urine would be delivered at body temperature, rather than a colder sample that might raise suspicions. Although these deceptions worked well, one thing that wasn't always controlled was the urine donor. There are stories of athletes testing positive for amphetamines on the Tour de France despite giving a false sample. Apparently the *soigneur* that was looking after them had just finished a long overnight drive and had needed something to keep him awake.

A more commercialised version of these methods came to prominence when a US football player was caught by airport security in 2005. Onterrio Smith was searched because he had suspicious vials of a white powder in his hand luggage. The good news for him was that the compound was not a banned drug; the bad news was that it was instead dried drug-free urine for filling the false penis he had brought along with him. Called *The Whizzinator* this is filled with the reconstituted dried urine and heated. Interestingly the manufacturers of this device, Puck Technology, fell foul not of the sporting authorities, but of the US government. The owners were found guilty of conspiring to defraud the administration of the federal workplace drug testing programme; the company's president was sent to prison.

In the case of sport it is clearly performance 'unenhancing' to compete with these devices actually fitted to your body. In the Tour de France teams would insert them prior to the drug tests. However, this became impossible once cyclists were marched straight off for testing at the end of the race. It is likely that the—admittedly invasive—regulations proved successful in reducing this particular form of cheating. Additionally urine is now stored for a long time and made available for later re-analysis; any foreign DNA can readily be detected. It is now a very high-risk move for an athlete to substitute someone else's urine for their own.

Even if they can't physically prevent giving a sample, athletes have sought chemical means to mask illegal drug use. While the simplest is to use a drug that is itself not tested for, there are other tricks that can be used. Athletes taking 'The Clear' (THG) were also taking 'The Cream'. This cream was a mixture of testosterone and epitestosterone designed to produce the same 1:1 ratio that the body produces. In this case you can increase testosterone levels without changing the T:TE ratio. As it is an abnormal T:TE ratio—not the absolute amount—that triggers a positive test, athletes are less likely to fail a test using the cream than testosterone alone.

Modifying the body's metabolism to alter the properties of a urine sample, such as its volume or acidity, can also make testing more difficult. Such compounds are called 'Masking agents'. They include probenecid, which can alter the excretion rate of anabolic steroids or diuretics which can dilute a urine sample and make drug detection more difficult. Shane Warne, the most successful Australian cricket bowler of all time, was banned for one year by the Australian Cricket Board when he tested positive for diuretics. There is no evidence that Warne, used performance enhancing drugs; indeed he vigorously denies this claiming that he took the pills to lose weight and improve his appearance. Nevertheless the penalties for being caught using a masking agent are generally as severe as those for using the drug it is designed to mask.

Perhaps the most spectacular coup by anti-doping agencies came when six members of the Finnish ski team were found guilty of taking hydroxy ethyl starch (HES) at the 2001 World Championships. HES is a plasma volume expander that dilutes blood samples and hence allows someone to appear to have a lower number of red blood cells. This can mask EPO and other blood doping offences, but in particular it ducked a rule that had just been introduced forbidding someone from competing when their red cell number was too high. HES is not a medical pill that could be taken by mistake. Instead it is a fluid that needs to be intravenously injected. The Finnish skiers knew exactly what they were doing; what they didn't know was that WADA had secretly introduced a test for HES. The positive results led to the International Ski Federation banning the skiers for two years.

Secrets and lies

Not surprisingly it is very rare to find an athlete admitting to doping in the absence of a positive test. Studies where the athlete remains anonymous can reveal this deception. One revealed that seven per cent of young elite athletes admitted to using doping substances, significantly more than the 0.8 per cent who failed random drug tests.[10] So it is no surprise that the first defence of any athlete accused of doping is denial. Confessions in the absence of a positive test generally require that the accusers have the full weight of law behind them. The fear of committing perjury—with the accompanying threat of a prison sentence—in front of a judicial inquiry or a US Grand Jury does occasionally loosen tongues.

When finally caught by a positive test most admit their guilt. So it is safe to assume that previous denials were lies. However, somewhat bizarrely some athletes who subsequently admit to a long history of doping insist that the specific test they got caught for was invalid. To this day Ben Johnson claims that he never took stanozolol, the steroid that resulted in his banning after the 100 m Olympic race in Seoul. Floyd Landis says he never took the steroids that resulted in the positive test and subsequent ban by the US Anti Doping Agency following his 2006 Tour de France victory. Landis claimed he was only on human growth hormone at the time.

A select few mount a spirited defence even after a positive test. Some of the excuses have the flavour of the naughty schoolboy who claimed that the dog ate his homework. Although the cyclist Frank Vandenbroucke suggested the only reason he had EPO in his possession in 2002 was to treat his dog's anaemia, he didn't escape a six month ban from the Belgian cycling federation. Some claim to have ingested the banned substance inadvertently. This was the case for the double Tour de France winner, Alberto Contador. He claimed that the banned substance clenbuterol found in a sample taken during the 2010 Tour was ingested from eating contaminated meat. Contador's explanation was ultimately not accepted by the Court of Arbitration for Sport. After a lengthy court case, the UCI and WADA appeal against the REFC ruling that cleared him in 2011 was upheld by the CAS in 2012. Contador continues to protest his innocence and has indicated that he is considering an appeal.

Some claim to have been the victims of conspiracy. Famously it was claimed that Justin Gatlin's USADA positive test for testosterone in 2006 was due to a cream rubbed in by a vindictive masseur; the 2004 Olympic 100m sprint champion still received a four year ban. In 1999 Dieter Baumann, the German middle distance runner, alleged that the positive test for the steroid nandrolone, for which he received a two year ban from the IAAF, was caused by someone spiking his toothpaste. The Russian 100 m hurdler Ludmilla Enquist claimed that her vitamin supplements were adulterated by a vindictive ex-husband leading to a 1993 positive steroid test and a four year ban from the IAAF. However, the Russian courts accepted her explanation and overturned Enquist's ban in 1995. The IAAF followed by reinstating her under its 'exceptional circumstances' rule, just in time for her to win the 1996 Olympic 100 m hurdles championship for her new adopted country Sweden. No such reprieve was available when she tested positive for steroids again in 2001 during her fledgling second career as a member of the Swedish bobsleigh team. She was banned for two years by the Swedish bobsleigh federation.

Sex is often used as an excuse. Some athletes just appear to be the victims of foolishness, like the Italian footballer Marco Borriello who applied his girlfriend's corticosteroid cream topically to treat his sexually transmitted disease and was banned for three months by the Italian FA. Others are more tragicomic. Olympic and world 400 m athletics champion LaShawn Merritt took a herbal male enhancement product ExtenZe. There is no evidence that ExtenZe enlarges penises and little evidence that the DHEA component enhances sports performance. Merritt's twenty-one month USADA ban in 2009 was for taking a product that was unlikely to affect his bedtime or athletic performance. Occasionally a case just seems sad; witness the ageing Japanese billiards champion Junsuke Inoue who took methyltestosterone in 1998, claiming he needed it to keep his wife satisfied in the bedroom. The Japanese Olympic committee immediately dropped him from their team for the Asian games. Testosterone is clearly no help in the billiard room and if Inoue had just waited a few months Viagra was about to be launched to revolutionise the field of sexual impotence.

The most famous sexual excuse was that of Dennis Mitchell, the US sprinter who claimed in 1998 that that his abnormally high positive test for testosterone was because he had five bottles of beer and sex four times as it was his wife's birthday and 'she deserved a treat'. Perhaps somewhat unusually the USA Track and Field authorities accepted this defence. It took the International Amateur Athletics Federation to restore a semblance of sanity to the proceedings noting that, while testosterone levels can rise after sex, the rise could in no way explain the amount found in Mitchell's system. The IAAF consequently reinstated the ban in 1999; Mitchell later admitted being injected with human growth hormone by the coach Trevor Graham involved with the BALCO scandal.[11] Despite the cynicism of some of the preceding paragraphs, I am sure that some athletes take compounds unaware that they contain banned substances. Sporting prowess does not always come with the wisdom to pick good friends and colleagues. Marion Jones, despite having three coaches, a husband and a boyfriend who were found to be involved in doping, still claims to have initially trusted they were giving her clean compounds.

However, whatever we might think of individual cases, it is clear that many athletes lie and make excuses in the full knowledge that they are taking drugs. It is difficult to see how an anti-doping policy based on drug testing can work without the strict liability rule that WADA enforces. If a compound is in your body, you are guilty. What is not clear is what WADA would do if a genuine case of malicious sabotage were discovered. It seems likely that this would mitigate the penalty, if not remove it entirely. This is presumably why Dieter Baumann offered a 100,000 Deutsche Mark reward in 2000 for anyone who could tell police how his toothpaste came to contain nandrolone.

Miscarriages of justice?

Sometimes athletes contest the science of the test itself. So how good does a drug test have to be to be foolproof? For many of the research results presented in this book, I have relied on the statistical idea that anything that happened at $P<0.05$ was a significant result i.e. if there was less than a 1 in 20 chance of something happening by chance it was a 'real' result. But these statistics refer to populations; they are not suitable for individuals. Just because a group on average runs faster after taking caffeine, it does not mean that every individual has run faster; in fact a small subpopulation might even have had a negative response and run slower.

So if I have developed a new test—let's say for gene doping—it would be no good to use average statistics and a probability cut-off of 1 in 20. Let's say this test was good 999 out of 1,000 times. This is much better than $P<0.05$. If the 1 in 1,000 chance is of missing a cheat (false negative) this might be acceptable. However, it is unacceptable if the 1 in 1,000 chance is of catching an innocent (false positive). This is because doping tests are not used to *confirm* whether someone is cheating; in this case a 1 in 1,000 error would probably be considered 'beyond reasonable doubt' by a jury. Instead they are the primary way of catching someone in the absence of any other evidence. A statistical calculation shows that for a 1 in a 1,000 chance of a false positive result, it would only take on average 693 tests to convict an innocent person.

As the aim is to test all athletes the chance of getting a false positive must be significantly less than the total number of random drug tests that will be carried out now or in the future. 1 in 1,000 is nowhere near this burden of proof. This shows the challenges in going from a laboratory result that is interesting to a validated test that will convict someone. This is also why the time from discovering a possible test to validating it for general use such that it can result in 'safe' convictions is frequently several years.

These examples are not merely theoretical. With more complex tests (such as those for EPO and HGH) athletes, their coaches and even other scientists have become more aggressive in querying positive test results. When athletes give a urine sample for testing, it is divided into two. If the A sample proves positive the B sample is tested for confirmation. In the vast majority of cases the B sample gives an identical result; if this is the case the athlete is found guilty. The EPO urine test is one of the few where the B sample has given a negative result after the A sample has been positive. Unfortunately for those doing the testing, those being reprieved included two high profile track runners, Marion Jones and Bernard Lagat. The EPO test is also one of the few where differences exist between laboratories. In 2008 the world famous Copenhagen Muscle Research Centre injected EPO into eight people.[12] They then passed the urine samples on to two WADA accredited laboratories. One WADA lab found all eight positives; the other labelled one negative and the other seven merely 'suspicious'; an athlete is not banned for a suspicious result.

So the urine test for EPO has the potential to yield false negatives. Can it also yield false positives? In 2005 the Belgian triathlete Rutger Beke was cleared despite failing both an A test and a B test. The reason? He was found to naturally excrete proteins following intense exercise that gave a positive EPO test; this led to his conviction being overturned. It seems that the antibody used in the WADA test is not completely specific for EPO;[13] it can bind to other proteins too. This can lead to a problem in a, admittedly unlikely, particular set of circumstances. First the athlete has a medical condition in which they excrete excess protein into the urine. If one of these proteins has a similar charge to EPO it will focus on the same place on the gel. If this protein can also react with the EPO antibody the conditions exist for a false positive test.

It should be noted that the test's inventor[14] and another WADA lab[15] both dispute the theory put forward by Beke's scientific allies. They claim that the fine details of the banding pattern should still enable discrimination between EPO and another protein. However, more than almost any other test the EPO test requires attention to the final result—there is a human judgement call on the banding pattern rather than result by automation. As stated by Martial Sugy—Head of the WADA accredited lab in Lausanne—'what is most important here is the experience of the eyes of the expert'. Two more triathletes were cleared following initial positive tests—the triathletes Virginia Berasategui and Ibán Rodríguez. Berasetegui and Rodriguez, as well as Lagat, have always denied doping. Subsequently, WADA have re-evaluated the rules for interpreting EPO tests. The EPO test is now one of the few that requires confirmation from two independent laboratories before a positive result is confirmed.

We shouldn't always rush to blame the scientists who devised a test if a result becomes difficult to interpret. One of the problems with testing athletes is that good quality control is required from start to finish. In many, possibly all, of the cases where false positives have been detected there has been mishandling of the sample. Bernard Lagats's urine sample was stored in the back of a car at almost 40°C. Lagat's urine became more alkaline, probably as a result of bacterial contamination—and this changed the pattern of proteins detected. The same bacterial contamination could have destroyed his natural EPO signal, making the anomalous signal stronger in comparison.

The most famous case of sample mistreatment was that of the unfortunate Diane Modahl. Over forty times the natural level of testosterone was found in Modahl's urine sample taken after an 800 m race in Portugal in June 1994. This resulted in her being sent home prior to competing in the Commonwealth Games that year; a consequent four-year ban from all competitions was imposed by the British Athletic Federation (BAF). Modahl has always protested her innocence. In 1995 her BAF ban was lifted, the IAAF following suit in 1996. To support her case she was able to show that her sample was left out on a bench in the Lisbon testing laboratory. Not being air conditioned, the sample sat in the Portuguese heat of 35°C for three days—ideal breeding temperatures for bacteria. Like Beke's sample Modahl's sample had become alkaline; it also showed no signs of the natural testosterone metabolites you would expect to see associated with the raised levels of testosterone.

The specific details of what happened in Modahl's sample are difficult to determine post hoc. It has not been possible to reproduce such a high testosterone signal simply by incubating urine at warm temperatures, although in 2002 a group at King's College London did show that yeast (*Candida*) contamination could convert some natural steroid precursors to testosterone.[16] Even then this was not at levels high enough to yield a positive doping test. We shall never know exactly what chemistry went on in Diane Modahl's sample, but the poor sample treatment left enough doubt in the test that the UK, and eventually the international, authorities subsequently rescinded her ban. She then sued the British Athletic Federation for £1 million claiming deliberate bias in her treatment. Although this action failed, the resulting court case ruined her and the associated costs were part of the reason that the federation went into administration in 1997. The Modahl case is a salutary warning of the personal and financial costs of making mistakes in doping tests.

The final example where tests have been called into question relates to the steroid nandrolone. In the early years of the twenty-first century the number of athletes failing tests due to nandrolone rose dramatically. In one sense this is not surprising; nandorolone is an effective anabolic steroid and seems to have fewer side effects than many others. Yet there seemed to be a mass of justifications from athletes, many of which led to bans being overturned.

The test for nandrolone is in fact a test for one of its metabolites 19-norandrosterone. This molecule is readily excreted from the body in urine and hence detectable by GC-MS. It is possible to have a natural level of 19-norandrosterone in your body without taking nandrolone. Therefore a limit is set above which it is assumed that you have been doping. The level at which this limit is set is crucial. Athletes claim that they can go over this limit by taking contaminated supplements or merely by eating meat products. Many professional tennis players on the world tour were not banned after positive nandrolone tests due to the suspicion that the metabolites produced in their body from the dietary supplements given to them by their own professional organisation were the source of the 19-norandrosterone found in their tests.

Contaminated meat has also been suggested to be the source of a positive test; the responses of the authorities have varied. In 1999

Petr Korda the Czech tennis player was banned for one year by the International Tennis Federation (ITF); the relatively light ban (initially it was only a fine) was due to the fact that the ITF accepted Korda's claim that he had no knowledge of how the nandrolone entered his system: one theory that has been suggested was that it arose from a meal of steroid-enhanced veal. In a similar vein, swimmers David Meca-Medina and Igor Majcen claimed that eating a stew of meat and offal from uncastrated boars for five consecutive days was the problem. While neither man's appeal was fully successful, the initial four-year ban imposed by the swimming authorities (FINA) was reduced to two years by the Court for Arbitration of Sport. The British bobsleigher Lenny Paul did not receive a ban at all, because the authorities accepted it was possible that the spaghetti bolognaise he ate had been contaminated due to the steroids fed to cattle by farmers.

In none of the cases where athletes have been reprieved was the supposed source of the nandrolone unequivocally proven to be the cause of the positive drug test; one is left with the suspicion that the different punishments could as much reflect the politics of the relevant organisations and the cost of litigation, as the specifics of the science itself. Litigation can be time consuming. The nandrolone case of the Spanish race walker Daniel Plaza took ten years to resolve until he was finally cleared of wrong doing by the Spanish Supreme Court.

The nandrolone situation is confusing and still far from resolved. Undoubtedly many—probably most—of the past and present positive tests are due to genuine steroid abuse. However, it is clear that some nutritional supplements can be converted into 19-norandrosterone in the body. The best scientific study of this phenomenon relates to the case of the UK athletes Mark Richardson and Dougie Walker. The discussion has at its core the setting of a reference range for a 'natural' level of steroid metabolites. The work led by Ron Maughan at Aberdeen University, showed that athletes who had failed a WADA nandrolone doping test in competition were able to create the same positive test in a controlled laboratory environment; they did this solely by combining vigorous exercise with supplements that were uncontaminated i.e. the supplements themselves passed the drug test.[17] The levels reached were five times the WADA limit. The only solution at present, and perhaps indefinitely, is for athletes to steer clear of all nutritional supplements.

This shouldn't be so difficult. First the evidence that supplements are needed for optimum performance is weak. But secondly, and more importantly, quality control in most of the companies selling these products is very poor. Many studies have shown that apparently innocent supplements contain ingredients that could cause an athlete to fail a doping test. In some cases this has been used in an athlete's defence. American swimmer Jessica Hardy had her two-year ban for clenbuterol reduced to one year by the American Arbitration Association when she produced evidence that the company told her the pills she was taking contained no untoward ingredients. In 2010 two South African rugby players, Chiliboy Ralepelle and Bjorn Basson, were initially banned by the South African Rugby Union for taking the stimulant methylhexaneamine. They seemed to be the victims of a bad batch of energy drink as the management had previously sent samples of the supplements to be tested with no positive results. The resultant ban was rescinded. The stimulant was contained in a supplement that their coaches gave to all the players before the game; Ralepelle and Basson were just the two unfortunate players tested on the day. Interestingly the South Africa team beat the Ireland team by a narrow margin 23–21; I suspect only the fact that it was a friendly match led to no official complaint from the Irish.

A new way forward?

If we perceive doping in sports as a war like the war on drugs, or the war on poverty, it is a war without end when it is fought in its present form. The dopers may think of an undetectable drug, the testers will find a way to detect it; the testers will find a new test, the dopers will switch to a new drug. And like Diane Modahl and Mark Richardson, there will be innocent casualties along the way. I see no way around this—it is an inevitable function of a policy that is based on drug tests and strict liability for athletes. Yet wars, even metaphorical ones, are worth fighting if the consequences of not fighting them are worse. Although the war on performance enhancing drugs cannot be won, it can be lost. So the question is whether it is worth fighting it at all?

This is a book designed to explore the science in order to allow people to come to their own decisions as to what actions should be implemented. As a biochemist my personal views are no more valid than those of any other sports fan. I am sure it is clear from reading between the lines in this book that if I were designing a WADA list of banned drugs and doping methods it would read somewhat differently to the present list. The list could certainly be cleaned up to remove items that are not related to sports performance. This would not only remove the recreational drugs, but also most over-the-counter medications. WADA and its scientists could then keep a clear focus on products that can potentially enhance performance; if a sport wished to fund its own workplace testing for recreational drugs—voluntary or otherwise—in order to protect its image or mange the health of it staff that could be done separately.

Yet changes of this type are still largely cosmetic. Are there really radical changes that could affect the rules of engagement between doper and tester? What about the idea suggested at the end of the last chapter of changing the doping regime to one based on the health of the athlete, rather than the purity of the sporting event itself? Could such a testing regime be introduced? In effect this would turn WADA into a medical agency. In reality, if the rules were changed so dramatically that athletes were not banned for doping there would be no need for an expensive testing regime at all. The normal health system could just as easily look after the health of athletes, dealing with them symptomatically as occurs now with recreational drug users.

Using the health of an athlete as the core of a test has been a feature of some recent developments in anti-doping science. The first example of such a policy is the health tests that were introduced in cycling in 1997. Before each race blood tests were introduced for cyclists to check that their blood red cell count was not too high. This was ostensibly marketed for the athletes' own safety; for if the red blood cell count were too high the blood could thicken and flow be restricted. The penalty for failure was more limited than for a doping offence. Of course there was a subtext in that—in the absence of EPO testing at the time—this was as good a way as any to level the playing field. The exact level of this test, high enough to deter cheats but not to catch innocent athletes using altitude training, is controversial. Somewhere between one to five per cent of drug free athletes have been suggested to have levels above the limits set.[18] The skier Eero Mantyranta with the mutation in his EPO receptor would never have been allowed to compete under these rules.

The penalties for failing these 'health' tests have mostly involved minor fines and a racing suspension until the red cell number comes down. Yet there is a clear social stigma involved. Many people perceive failing this 'health' test as evidence of doping. The 1998 Tour de France winner Marco Pantini never fully recovered mentally from failing a red cell test in the Tour of Italy in 1999. UCI tests showed that the British cyclist Rob Hayles had a red blood cell level 0.3% above that permitted and had to withdraw from the 2008 World Championships. Hayles has always denied doping. Red cell levels can fluctuate naturally and with the extent of training; Hayles was passed fit to compete after extensive testing over the subsequent two weeks showed no signs of doping. However, any test that results in someone being banned from competition cannot by definition be confidential and so—even if not judged by sporting bodies—Hayles still had to defend himself in the media.

Health checks are not limited to cycling and cross-country skiing. Paralympic athletes with severe spinal cord injury cannot raise their heart rate above about 130 beats per minute. Although their upper body may be fully functional the lack of a global response to exercise, including production of adrenaline, limits their cardiac output and hence performance. The solution is to induce a condition called autonomic dysreflexia (AD). Colloquially termed 'boosting' this is the triggering of neural responses below the spinal cord lesion that send 'stress' messages back to the brain. This resulting increase in maximal heart rate can improve performance. How do you activate the nervous system below the spinal column when you have no conscious control of this part of the body? A variety of methods are used to induce the necessary stress—these include excessive tightening of leg straps, blocking a urinary catheter to over-fill the bladder, sitting on the scrotum or even in extreme cases breaking a bone. Of course what seems extreme is a matter of perception—the athlete feels no pain after all. Nevertheless inducing AD is banned.

Lest anyone think that attitudes towards banned methods differ between Olympic and Paralympic sport, a recent anonymous questionnaire found that fifteen per cent of male athletes with serious spinal cord injury admitted to using AD to enhance their performance.[19] Paralympians are just as performance obsessed, and no more idealistic, than normal Olympians. It is impractical to perform a complete physical inspection of all athletes to check for broken toes or blocked catheters. So the only current test is to measure the resting blood pressure. Following AD this will be high; in fact in some cases so high that this method is potentially life threatening. Just as is the case with cyclists with too many red blood cells, this blood pressure test is used to remove people from the immediate competition for their health benefit. However, it is not a very specific test for AD. Even if a person's 'normal' resting blood pressure were known, that number would vary from day-to-day and even hour-to-hour. Consequently, a single blood pressure reading cannot be used as the sole means of banning someone.

There is a fundamental difference between these tests and the historical activities of an anti-doping laboratory. The new tests are trying to monitor the normal functioning of the athlete's body, rather than looking for the drug directly in the urine or blood sample. Monitoring metabolism and physiology is at the heart of many new tests. The EPO and HGH urine tests have been notoriously poor at detecting abuse even a few days after athletes have stopped taking drugs. However, there are longer-term impacts on the body's physiology. For example there is a blood test for EPO that looks for biological markers that suggest recent increases in new red cell production. These include the number of immature red cells and indications that iron metabolism is being upregulated to provide the necessary iron to enable the haemoglobin protein to bind oxygen. These markers are then seen as a 'smoking gun' for EPO abuse.[20]

Similarly an HGH blood test blood has been under development that measures molecules that respond to an increase in HGH. These are the insulin-like growth factor 1 (IGF-1) that drives protein synthesis and a molecule called P3NP that is a byproduct of collagen, the molecule that supports the structure of new cells.[21] The increases in these molecules last longer in the body than the original HGH; they extend the duration of detection from one day to almost two weeks, long enough to significantly increase the chance of success in a random drug test.

The tests for HGH and EPO have two things in common. They are both blood tests and they are indirect. While athletes in Olympic sports have agreed—or been forced to agree—to blood tests in competition, it is harder and costlier to implement these in worldwide random testing. It is also much harder to introduce a blood test in a sport that does not have to bow to WADA's dictats. Professional sports in the USA with strong player labour unions—such as ice hockey, baseball and football—have proved especially resistant to such testing.

There is also a scientific issue here. An indirect test is a statistical test. It requires that we know the baseline values for all individuals. Whilst there are some validated statistical tests for normal body compounds—the abnormal testosterone/epistesterone (T:TE) ratio being one—an athlete can always claim to be a physiological outlier. Validation is therefore complicated and costly. The problem lies in the statistics. There is a large variation between people in their normal levels of these biological markers. So trying to set a limit for a single biomarker that is high enough to catch a doper is almost certainly going to cause many false positives. Combining biomarkers reduces the false positives—hence the success of the T:TE ratio.

The variation of the levels of a marker such as IGF-1 within one individual over time is much lower than the variation between many individuals at any one specific time: in statistical terms the 'within subject' variability is much smaller than the 'between subject' variability. So these indirect tests become much more powerful if we have baseline numbers for an athlete that we can use for a personalised comparison. Enter the biological passport. An athlete would submit to a test when they are known to be drug free. Blood and urine samples would be monitored over time and the results entered in their passport; anomalous changes due to doping could then be picked up.

Such tests need not be restricted to those we have already mentioned. The big move forward in science in the last decade has been the development of rapid, and cheap methods of measuring large amounts of individualised biological data. With simple tests we can now measure the total amount of DNA in the body (genomics); the total amount of RNA currently being transcribed from that DNA (transcriptomics); the resulting total complement of protein produced from the RNA (proteomics) and the effects on the other molecules in the body (metabolomics). These 'omic' techniques—the simultaneous measurement of everything—have revolutionised biology; they will ultimately impact on drug testing. Recently I was speaking to a German biophysicist who used to play professional basketball. She was optimistic at the prospect that these tests would finally make sport drug free. However, caution is warranted; as we know it is a very long way from science that shows differences in a laboratory to a validated test that will stand up in court. We should bear in mind that the HGH blood test grew out of a project targeting the Sydney 2000 Olympics called GH2000; it was still not in use by 2010.

While I have drawn extensively on the parallels between doping in sport and drug use in society, there are some obvious differences. Chiefly, the number of elite sportspeople is tiny compared to the number of recreational drug users. It is notable that much of the success in recent years at combating drug use can be attributed to the involvement of police rather than drug testing laboratories. Their job is made easier as there are far fewer suppliers and users in sports doping than in recreational drug use, at least at the elite level.

The relatively small number of elite athletes also means that it would be realistically possible to test on a monthly basis all athletes that would play football in the World Cup, play in a Super Bowl or compete in the Olympic games. No clean passport—no competition. Of course this would require massive political will and cooperation between governments. However, it is technically feasible to implement if someone had the vision and the money.

In the case of cycling, biological passports have already been introduced. The Science and Industry Against Blood doping (SIAB) research consortium promoted the use of a 'Haematologic Passport' where the athlete's normal blood values (both mature and immature red cells) are used to compare changes observed before and during events. Union Cycliste Internationale uses an athlete passport system based on these ideas to monitor almost 1,000 elite cyclists for blood doping at a cost of €20 million. It is envisiged that this will ultimately be extended to hormone tests that would indicate steroid or HGH abuse. Of course extending passports to all levels of sport would be prohibitively expensive. In the case of blood doping this doesn't matter. The benefits are immediate and transient and a passport system for current athletes should work fine. However, in the case of steroids, muscle growth in the athlete's youth could have long-term benefits that are retained even after the treatment is stopped. Passports could only be part of a solution here.

Even passports are not without their critics. In May 2010 the Union Cycliste Internationale (UCI) charged three riders solely on the basis of passport anomalies. However, the most high profile victim, Franco Pellizotti—winner of the 'King of the Mountains' at the 2009 Tour de France—had his two-year UCI ban overturned by the Italian courts. Although he had three anomalous readings, the court was not convinced that on their own they were sufficient evidence of doping.

Unfortunately as each athlete's readings are kept secret it is not possible for an outsider to know why the committee recommended the initial ban, or why it was overturned. In the case of Pellizotti the case went to the Court for Arbitration for Sport in March 2011. Despite his protestations of innocence, the court reinstated the UCI two-year ban; Pellizotti was fined €115,000 and lost his 2009 'King of the Mountains' title.

If the system is going to work in the future and not be continually mired in the courts it seems to me that complete transparency and openness is necessary for the passports, even at the risk of allowing the dopers to have a better idea of how to avoid suspicion. Why are the cycling authorities wary of such openness? Well there is evidence from the confessions of the cyclist Floyd Landis that cyclists have modified their EPO doping regime in response to the introduction of passports. The trick to fool the current biological passport seems to be to dope little and often, rather than in the one large dose that leads to suspicious spikes in blood parameters. The war continues.

Concluding Remarks

'Everyone is entitled to their own opinion, but not their own facts.'
Senator Daniel Moynihan

Most books about drugs in sport are not written by scientists. Indeed in some books scientists only feature as the villains in the story. Paul Dimeo argues that scientists not only designed the drug tests, but also played a major role in forcing their implementation on athletes. In his words, however well-intentioned they might have been, 'scientists . . . put forward a fundamentalist view of ethics and a pseudo-imperialist strategy of moral diffusion' (chapter 9 ref. 1). A. J. Schneider goes even further to argue that the only ethical approach is for the athletes themselves—not scientists or politicians—to determine and manage doping policy.[1]

This book has sought to redress the imbalance. It is clear that the scientific method can offer answers to the factual questions about human performance. However, science can do a little more. When it comes to making practical and ethical policy there is a devil in the scientific detail (or 'facts' if you like) that is absolutely required if we are to make informed moral and political choices. It matters whether human growth hormone is of less value than creatine in improving human performance; or that caffeine outperforms amphetamines. These scientific questions inform which drugs should be banned and how limited resources should best be spent in testing and enforcement.

Science also points to the future of doping. It tells us that the natural mutations that enhance a Usain Bolt or a Paula Radcliffe are more important than any pharmaceutical enhancement that could be given to a less genetically endowed athlete; that the real power of gene doping requires changing the nature of foetuses not adults; that in the long term designing a superhuman for sport is the same ethical question as designing a more intelligent baby or even one with a certain eye colour.

Yet, as I hope this book makes clear, however we make use of science, we can no more 'win' a war on drugs in sport than we can 'win' a war on drugs in society. There are many performance-enhancing pharmaceuticals that remain to be discovered. That this knowledge is largely unexploited is due to the limited spending on sport compared with medical or military research.

However, some things can be done. The testing regime can be based on more scientific principles. I worry that sport has become involved in the morality of drug use by banning, and testing for, recreational drugs. I also feel deeply the pain of athletes like Andreea Răducan or Alan Baxter, who lost their Olympic dreams over the use of a minor cold remedy. Yet I am troubled about the consequences of a drugs free-for-all. All parents would surely be dismayed by the phone call that the US geneticist Lee Sweeney received from a high school coach asking him to gene dope his whole football team. We should not underestimate the power of science unrestricted by moral ambiguity.

There are no simple solutions, but improvements can be made to enhance fairness. In world sport we have the advantage of a single organisation that controls all doping regulations. Let's use this power wisely, and informed by science to make sport that is as fair as possible to both athletes and fans.

Notes and References

PREFACE

1. United States Anti-Doping Agency (2012) U.S. Postal Service Pro Cycling Team Investigation. **Available from:** *http://cyclinginvestigation.usada.org*
2. J. T. Dalton, K. G. Barnette, C. E. Bohl, M. L. Hancock, D. Rodriguez, S. T. Dodson, R. A. Morton, and M. S. Steiner (2011) The selective androgen receptor modulator GTx-024 (enobosarm) improves lean body mass and physical function in healthy elderly men and postmenopausal women: results of a double-blind, placebo-controlled phase II trial, *J. Cachexia Sarcopenia Muscle* 2, 153–161.
3. FibroGen Inc (2012) FibroGen announces results showing that FG-4592, an oral hypoxia-inducible factor prolyl hydroxylase inhibitor (HIF-PHI), corrects anemia without intravenous iron supplementation in incident dialysis patients. **Available from:** *http://www.fibrogen.com/press/release/pr_1352127487*

PROLOGUE

1. 'The most corrupt race ever', *Observer Sport Monthly*, Sunday August 1, London, 2004.
2. O. Slot, 'The Athlete: Calvin Smith,' *The Times, September* 24, 2003. http://www.thetimes.co.uk/tto/sport/athletics/article2375062.ece
3. L. Williams, 'Sprinters testify against coach in S.F. trial,' *San Fransisco Chronicle, May* 23, 2008. http://articles.sfgate.com/2008-05-23/news/17156398_1_trevor-graham-sprint-capitol-drug-connection
4. C. L. Dubin, *Commission of Inquiry into the Use of Drugs and Banned Practices Intended to Increase Athletic Performance.* (Ottawa, Canadian Publishing Centre, 1990).
5. L. Banack, 'Adjudication: Appeal by Desai Williams of a life time withdrawal by Sport Canada to access to direct federal funding.' *Sport Canada* (2010).

6. Canadian Broadcasting Company, 'Damning evidence in the Dubin Inquiry' (October 3, 1989) http://archives.cbc.ca/sports/drugs_sports/clips/8964/ *Accessed February 22, 2011.*
7. T. Fordyce, British Broadcasting Company, 'Ceplak rejects cheat claim' (August 9 2002) http://news.bbc.co.uk/sport1/hi/athletics/specials/european_athletics/2182003.stm *Accessed February 22, 2011.*
8. *Daily Mail* (UK), 'Dame Kelly insists: Drug testing is now a vital way of life' http://www.dailymail.co.uk/sport/othersports/article-412761/Dame-Kelly-insists-Drug-testing-vital-way-life.html *Accessed February 22, 2011.*
9. B. Goldman, P. J. Bush, R. Klatz, *Death in the locker room.* (Century, London, 1984).

CHAPTER 1

1. R. Renson, Fair play: its origins and meanings in sport and society. *Kinesiology* **41**, 5–18 (2009).
2. W. Voet, *Breaking the Chain.* (Yellow Jersey Press, London, 2001).
3. United States v. Coca Cola Co. of Atlanta. *United States Reports* 41 (1916).
4. Philostratos, *Philostratos uber Gymnastik. Sammlung wissenschaftlicher Kommentdare zu griechischen und romischen Schriftsteltern (translated by J. Juthner).* (B.R. Gruner, Amsterdam, 1969).
5. A. Curry, The Gladiator Diet. *Archaeology* **61**, (November/December, 2008).
6. N. Szczepanik, *The Times* (UK), 'Are sport and sex a recipe for success?' (September 24, 2009) http://www.timesonline.co.uk/tol/sport/more_sport/article6847419.ece *Accessed February 22, 2011.*
7. B. Joy, *Forward Arsenal: The Arsenal Story, 1888–1952.* (Phoenix House, London, 1952).
8. W. W. Franke, B. Berendonk, 'Hormonal doping and androgenization of athletes: a secret program of the German Democratic Republic government'. *Clin. Chem.* **43**, 1262–79 (1997).
9. V. Conte, Letter to Dwain Chambers (16 May, 2008) http://news.bbc.co.uk/sport1/hi/olympics/athletics/7403158.stm *Accessed Feb 22, 2011.*
10. R. M. Pirsig, *Zen and the Art of Motorcycle Maintenance: An Inquiry into Values.* (William Morrow & Company, New York, 1974).
11. D. M. Bramble, D. E. Lieberman, 'Endurance running and the evolution of Homo'. *Nature* **432**, 345–52 (2004).
12. B. Carey, *LiveScience*, 'Scientists Build "Frankenstein" Neanderthal Skeleton' (10 March 2005) http://www.livescience.com/history/050310_neanderthal_reconstruction.html *Accessed Sept. 28, 2009.*

13. U. Segerstrale, *Defenders of the Truth: The Battle for Science in the Sociobiology Debate and Beyond*. (Oxford University Press, Oxford, 2000).

14. A. Kessel, *The Observer* (UK), 'Unattainable records leave female athletes struggling for acclaim' (16 August, 2009) http://www.guardian.co.uk/sport/blog/2009/aug/16/world-athletics-championships-records *Accessed February* 22, 2011.

15. *Der Spiegel* (International), 'How Dora the Man Competed in the Woman's High Jump' (September 15 2009) http://www.spiegel.de/international/germany/0,1518,649104,00.html *Accessed February* 22, 2011.

16. J. L. Simpson et al., 'Gender verification in the Olympics'. *JAMA* **284**, 1568–9 (2000).

17. *The Daily Telegraph*, July 06 2010. http://www.telegraph.co.uk/sport/othersports/athletics/7873921/Caster-Semenya-anatomy-of-her-case.html.

18. A. de la Chapelle, A. L. Traskelin, E. Juvonen, 'Truncated erythropoietin receptor causes dominantly inherited benign human erythrocytosis'. *Proc. Natl. Acad. Sci. U S A* **90**, 4495–9 (1993).

CHAPTER 2

1. N. P. Linthorne, 'Was Flojo's 100-m world record wind-assisted?' *Track Technique* **127**, 4052–3 (1994).

2. C. T. Davies, 'Effects of wind assistance and resistance on the forward motion of a runner'. *J. Appl. Physiol.* **48**, 702–9 (1980).

3. P. Watson et al., 'Acute dopamine/noradrenaline reuptake inhibition enhances human exercise performance in warm, but not temperate conditions'. *J. Physiol.* **565**, 873–83 (2005).

4. J. M. Carre, 'No place like home: testosterone responses to victory depend on game location'. *Am. J. Hum. Biol.* **21**, 392–4 (2009).

5. R. Tucker, 'The Science of Sport, The Four-minute mile: The value of integration of physiology and mental aspects of performance' (January 14, 2009) http://www.sportsscientists.com/2009/01/mind-vs-matter.html *Accessed February* 22, 2011.

6. R. Beneke, M. J. Taylor, 'What gives Bolt the edge—A.V. Hill knew it already!' *J Biomech.* **43**, 2241–3 (2010).

7. J. L. Ray, 'The Top 10 Baseball Teams of All Time' (March 24, 2007) http://www.suite101.com/content/baseballs-ten-best-teams-ever-a17072 *Accessed February* 22, 2011.

8. T. Reilly, V. Thomas, 'A time motion analysis of work rate in different positional roles in professional match play'. *J. Hum. Mov. Stud.* **2**, 87–99 (1976).

9. J. Bangsbo, 'Energy demands in competitive soccer'. *J. Sports Sci.* **12**, S5–S12 (1994).

10. B. Tindall, *Guardian* (UK), 'The drugs do work, Nigel' (Wednesday 21 May, 2008) http://www.guardian.co.uk/music/musicblog/2008/may/21/blairtindallwedsampic *Accessed February* 22, 2011.

11. G. A. Gates et al., 'Effect of beta blockade on singing performance'. *Ann. Otol.Rhinol. Laryngol.* **94**, 570–4 (1985).

CHAPTER 3

1. T. Noakes, *Lore of Running*. 4th edn, (Oxford University Press Southern Africa, 2001).

2. C. Foster, D. L. Costill, W. J. Fink, 'Effects of preexercise feedings on endurance performance'. *Med. Sci. Sports* **11**, 1–5 (1979).

3. W. M. Sherman, M. C. Peden, D. A. Wright, 'Carbohydrate feedings 1 h before exercise improves cycling performance'. *Am. J. Clin. Nutr.* **54**, 866–70 (1991).

4. J. Bergstrom, E. Hultman, 'The effect of exercise on muscle glycogen and electrolytes in normals'. *Scand. J. Clin. Lab. Invest.* **18**, 16–20 (1966).

5. J. Bergstrom, E. Hultman, 'Synthesis of muscle glycogen in man after glucose and fructose infusion'. *Acta Med. Scand.* **182**, 93–107 (1967).

6. W. M. Sherman, D. L. Costill, W. J. Fink, J. M. Miller, 'Effect of exercise-diet manipulation on muscle glycogen and its subsequent utilization during performance'. *Int. J. Sports Med.* **2**, 114–18 (1981).

7. C. W. Nicholas, C. Williams, H. K. Lakomy, G. Phillips, A. Nowitz, 'Influence of ingesting a carbohydrate-electrolyte solution on endurance capacity during intermittent, high-intensity shuttle running'. *J. Sports Sci.* **13**, 283–90 (1995).

8. B. Simi, B. Sempore, M. H. Mayet, R. J. Favier, 'Additive effects of training and high-fat diet on energy metabolism during exercise'. *J. Appl. Physiol.* **71**, 197–203 (1991).

9. S. D. Phinney, B. R. Bistrian, W. J. Evans, E. Gervino, G. L. Blackburn, 'The human metabolic response to chronic ketosis without caloric restriction: preservation of submaximal exercise capability with reduced carbohydrate oxidation'. *Metabolism* **32**, 769–76 (1983).

10. L. M. Burke, 'Fueling strategies to optimize performance: training high or training low?' *Scand J Med Sci Sports* **20** Suppl 2, 48–58 (2010).

11. E. S. Chambers, M. W. Bridge, D. A. Jones, 'Carbohydrate sensing in the human mouth: effects on exercise performance and brain activity'. *J Physiol* **587**, 1779–94 (2009).

12. P. Hespel, W. Derave, 'Ergogenic effects of creatine in sports and rehabilitation'. *Subcell. Biochem.* **46**, 245–59 (2007).

13. E. P. Brass, 'Supplemental carnitine and exercise'. *Am J Clin Nutr* **72**, 618S–23S (2000).

14. G. Jones, 'Caffeine and other sympathomimetic stimulants: modes of action and effects on sports performance'. *Essays Biochem* **44**, 109–23 (2008).

15. H. Westerblad, D. G. Allen, J. Lannergren, 'Muscle fatigue: lactic acid or inorganic phosphate the major cause?' *News Physiol Sci* **17**, 17–21 (2002).

16. V. Ööpik, S. Timpmann, K. Kadak, L. Medijainen, K. Karelson, 'The effects of sodium citrate ingestion on metabolism and 1500 m racing time in trained female runners'. *J Sports Sci. & Med.* **7**, 125–31 (2008).

17. L. L. Spriet, C. G. Perry, J. L. Talanian, 'Legal pre-event nutritional supplements to assist energy metabolism'. *Essays Biochem* **44**, 27–43 (2008).

18. C. Nicholas, 'Legal nutritional supplements during a sporting event'. *Essays Biochem.* **44**, 45–61 (2008).

19. E. A. Newsholme, E. Blomstrand, 'Branched-chain amino acids and central fatigue'. *J. Nutr.* **136**, 274S–6S (2006).

20. J. L. Pannier, J. J. Bouckaert, R. A. Lefebvre, 'The antiserotonin agent pizotifen does not increase endurance performance in humans'. *Eur J Appl Physiol Occup Physiol* **72**, 175–8 (1995).

21. C. Duncan et al., 'Chemical generation of nitric oxide in the mouth from the enterosalivary circulation of dietary nitrate'. *Nat. Med.* **1**, 546–51 (1995).

22. F. J. Larsen, E. Weitzberg, J. O. Lundberg, B. Ekblom, 'Effects of dietary nitrate on oxygen cost during exercise'. *Acta Physiol. (Oxf)* **191**, 59–66 (2007).

23. S. J. Bailey et al., 'Dietary nitrate supplementation reduces the O2 cost of low-intensity exercise and enhances tolerance to high-intensity exercise in humans'. *J. Appl. Physiol.* **107**, 1144–55 (2009).

CHAPTER 4

1. P. D. Wagner, 'Counterpoint: in health and in normoxic environment VO2max is limited primarily by cardiac output and locomotor muscle blood flow'. *J. Appl. Physiol.* **100**, 745–7 (2006).

2. B. Ekblom, G. Wilson, P. O. Astrand, 'Central circulation during exercise after venesection and reinfusion of red blood cells'. *J. Appl. Physiol.* **40**, 379–83 (1976).

3. A. J. Brien, T. L. Simon, 'The effects of red blood cell infusion on 10 km race time'. *Jama* **257**, 2761–5 (1987).

4. G. Hopfl, O. Ogunshola, M. Gassmann, 'Hypoxia and high altitude. The molecular response'. *Adv. Exp. Med. Biol.* **543**, 89–115 (2003).

5. B. D. Levine, J. Stray-Gundersen, 'Point: positive effects of intermittent hypoxia (live high:train low) on exercise performance are mediated primarily by augmented red cell volume'. *J. Appl. Physiol.* **99**, 2053–5 (2005).

6. F. Celsing, J. Svedenhag, P. Pihlstedt, B. Ekblom, 'Effects of anaemia and stepwise-induced polycythaemia on maximal aerobic power in individuals with high and low haemoglobin concentrations'. *Acta Physiol. Scand.* **129**, 47–54 (1987).

7. L. C. Clark, Jr, F. Gollan, Survival of mammals breathing organic liquids equilibrated with oxygen at atmospheric pressure. *Science* **152**, 1755–6 (1966).

8. G. S. Hughes, Jr et al., 'Hemoglobin-based oxygen carrier preserves submaximal exercise capacity in humans'. *Clin. Pharmacol. Ther.* **58**, 434–43 (1995).

9. C. E. Cooper, 'Radical producing and consuming reactions of hemoglobin: how can we limit toxicity?' *Artif. Organs* **33**, 110–14 (2009).

10. T. A. Silverman, R. B. Weiskopf, 'Hemoglobin-based oxygen carriers: current status and future directions'. *Transfusion* **49**, 2495–515 (2009).

11. L. Douay, H. Lapillonne, A. G. Turhan, 'Stem cells—a source of adult red blood cells for transfusion purposes: present and future'. *Crit. Care Clin.* **25**, 383–98 (2009).

12. J. A. Calbet et al., 'Maximal muscular vascular conductances during whole body upright exercise in humans'. *J. Physiol.* **558**, 319–31 (2004).

13. R. S. Richardson, K. Tagore, L. J. Haseler, M. Jordan, P. D. Wagner, 'Increased VO2 max with right-shifted Hb-O2 dissociation curve at a constant O2 delivery in dog muscle in situ'. *J. Appl. Physiol.* **84**, 995–1002 (1998).

14. R. M. Winslow, 'MP4, a new nonvasoactive polyethylene glycol-hemoglobin conjugate'. *Artif. Organs* **28**, 800–6 (2004).

15. S. Leigh-Smith, 'Blood boosting'. *Br. J. Sports Med.* **38**, 99–101 (2004).

16. S. George, B. Haake, (ed.), 'Jaksche admits taking banned substances & blood doping' (July 1, 2007) http://autobus.cyclingnews.com/news.php?id = news/2007/jul07/jul01news *Accessed 26 February, 2011.*

17. N. S. Kenneth, S. Rocha, 'Regulation of gene expression by hypoxia'. *Biochem. J.* **414**, 19–29 (2008).

18. M. Y. Koh, T. R. Spivak-Kroizman, G. Powis, 'HIF-1alpha and cancer therapy'. *Recent Results Cancer Res* **180**, 15–34 (2010).

CHAPTER 5

1. J. Antonio, W. J. Gonyea, 'Skeletal muscle fiber hyperplasia'. *Med. Sci. Sports Exerc.* **25**, 1333–45 (1993).

2. A. L. Goldberg, J. D. Etlinger, D. F. Goldspink, C. Jablecki, 'Mechanism of work-induced hypertrophy of skeletal muscle'. *Med. Sci. Sports* **7**, 185–98 (1975).

3. H. Wackerhage, A. Ratkevicius, 'Signal transduction pathways that regulate muscle growth'. *Essays Biochem.* **44**, 99–108 (2008).

4. A. C. McPherron, A. M. Lawler, S. J. Lee, 'Regulation of skeletal muscle mass in mice by a new TGF-beta superfamily member'. *Nature* **387**, 83–90 (1997).

5. A. C. McPherron, S. J. Lee, 'Double muscling in cattle due to mutations in the myostatin gene'. *Proc. Natl. Acad. Sci. U S A* **94**, 12457–61 (1997).

6. M. Schuelke et al., 'Myostatin mutation associated with gross muscle hypertrophy in a child'. *N. Engl. J. Med.* **350**, 2682–8 (2004).

7. K. D. Tipton, A. A. Ferrando, 'Improving muscle mass: response of muscle metabolism to exercise, nutrition and anabolic agents'. *Essays Biochem.* **44**, 85–98 (2008).

8. H. Fouillet et al., 'Absorption kinetics are a key factor regulating postprandial protein metabolism in response to qualitative and quantitative variations in protein intake'. *Am. J. Physiol. Regul. Integr. Comp. Physiol.* **297**, R1691–R1705 (2009).

9. T. M. Robinson, D. A. Sewell, P. L. Greenhaff, 'L-arginine ingestion after rest and exercise: effects on glucose disposal'. *Med. Sc.i Sports Exerc.* **35**, 1309–15 (2003).

10. G. M. Fogelholm, H. K. Naveri, K. T. Kiilavuori, M. H. Harkonen, 'Low-dose amino acid supplementation: no effects on serum human growth hormone and insulin in male weightlifters'. *Int. J. Sport Nutr.* **3**, 290–7 (1993).

11. R. J. Louard, E. J. Barrett, R. A. Gelfand, 'Effect of infused branched-chain amino acids on muscle and whole-body amino acid metabolism in man'. *Clin. Sci. (Lond)* **79**, 457–66 (1990).

12. G. J. Wilson, J. M. Wilson, A. H. Manninen, 'Effects of beta-hydroxy-beta-methylbutyrate (HMB) on exercise performance and body composition across varying levels of age, sex, and training experience: A review'. *Nutr. Metab. (Lond)* **5**, 1 (2008).

13. J. S. Volek, E. S. Rawson, 'Scientific basis and practical aspects of creatine supplementation for athletes'. *Nutrition* **20**, 609–14 (2004).

14. E. Louis, U. Raue, Y. Yang, B. Jemiolo, S. Trappe, 'Time course of proteolytic, cytokine, and myostatin gene expression after acute exercise in human skeletal muscle'. *J. Appl. Physiol.* **103**, 1744–51 (2007).

15. D. S. Willoughby, 'Effects of heavy resistance training on myostatin mRNA and protein expression'. *Med. Sci. Sports Exerc.* **36**, 574–82 (2004).

16. A. Saremi et al., 'Effects of oral creatine and resistance training on serum myostatin and GASP-1'. *Mol. Cell Endocrinol.* **317**, 25–30 (2010).

17. M. N. Fedoruk, J. L. Rupert, 'Myostatin inhibition: a potential performance enhancement strategy?' *Scand. J. Med. Sci. Sports* **18**, 123–31 (2008).

CHAPTER 6

1. Y. Wu et al., 'Identification of androgen response elements in the insulin-like growth factor I upstream promoter'. *Endocrinology* **148**, 2984–93 (2007).

2. Aristotle, *The Works of Aristotle, trans. Anon.: The problems of Alexander Aphrodiseus.* J. Manis, ed., Pennsylvania State University , Electronic Classics Series. (Hazleton, Pa., 2005).

3. E. R. Freeman, D. A. Bloom, E. J. McGuire, 'A brief history of testosterone'. *J. Urol.* **165**, 371–3 (2001).

4. I. Berlin (1925) 'Monkee Doodle-Do'(song).

5. E. Simonson, W. M. Kearns, E. N, 'Effect of Methyl Testosterone Treatment on Muscular Performance and the Central Nervous System of Older Men'. *J Clin. Endocrinol.* **4**, 528–34 (1944).

6. W. P. VanHelder, E. Kofman, M. S. Tremblay, 'Anabolic steroids in sport'. *Can J Sport Sci* **16**, 248–57 (1991).

7. S. Bhasin et al., 'The effects of supraphysiologic doses of testosterone on muscle size and strength in normal men'. *N. Engl. J. Med.* **335**, 1–7 (1996).

8. A. A. Ferrando et al., 'Testosterone injection stimulates net protein synthesis but not tissue amino acid transport'. *Am. J. Physiol.* **275**, E864–71 (1998).

9. A. A. Ferrando, M. Sheffield-Moore, D. Paddon-Jones, R. R. Wolfe, R. J. Urban, 'Differential anabolic effects of testosterone and amino acid feeding in older men'. *J. Clin. Endocrinol. Metab.* **88**, 358–62 (2003).

10. A. T. Kicman et al., 'Effect of androstenedione ingestion on plasma testosterone in young women; a dietary supplement with potential health risks'. *Clin. Chem.* **49**, 167–9 (2003).

11. M. S. Schmidt, 'Manny Ramirez is banned,' *New York Times*, May 7, 2009. www.nytimes.com/2009/05/08/sports/baseball/08ramirez.html

12. M. Fainaru-Wada, T. J. Quinn, ESPN.comSources: Ramirez used fertility drug (May 8, 2009) http://sports.espn.go.com/mlb/news/story?id=4148907 *Accessed 3/12/2011.*

13. S. Doessing et al., 'Growth hormone stimulates the collagen synthesis in human tendon and skeletal muscle without affecting myofibrillar protein synthesis'. *J. Physiol.* **588**, 341–51 (2010).

14. M. J. Rennie, 'Claims for the anabolic effects of growth hormone: a case of the emperor's new clothes?' *Br. J. Sports Med.* **37**, 100–5 (2003).

15. BioGrid 3.1 (Biological General Repository for Interaction Datasets) http://thebiogrid.org/ *Accessed Search for NR3C4, April 2, 2011.*

16. T. Cook, W. P. Sheridan, 'Development of GnRH antagonists for prostate cancer: new approaches to treatment'. *Oncologist* **5**, 162–8 (2000).

CHAPTER 7

1. J. R. Docherty, 'Pharmacology of stimulants prohibited by the World Anti-Doping Agency (WADA)'. *Br. J. Pharmacol.* **154**, 606–22 (2008).

2. S. Freud, 'Ueber Coca'. *Centrabl f d ges Therapie (Wien)* **2**, 289–314 (1884).

3. S. Freud, 'A contribution to the knowledge of the effect of cocaine', *Vienna Medical Weekly*, 1885 in *Cocaine Papers*, R. Byck, ed. (Stonehill Publishing Co, New York, NY, 1974).

4. K. C. Berridge, 'The debate over dopamine's role in reward: the case for incentive salience'. *Psychopharmacology (Berl)* **191**, 391–431 (2007).

5. E. Davis, R. Loiacono, R. J. Summers, 'The rush to adrenaline: drugs in sport acting on the beta-adrenergic system'. *Br. J. Pharmacol.* **154**, 584–97 (2008).

6. K. D. Fitch et al., 'Asthma and the elite athlete: summary of the International Olympic Committee's consensus conference, Lausanne, Switzerland, January 22–24, 2008'. *J. Allergy Clin. Immunol.* **122**, 254–60, 260 e251–257 (2008).

7. J. M. Weiler, E. J. Ryan, 3rd, 'Asthma in United States olympic athletes who participated in the 1998 olympic winter games'. *J. Allergy Clin. Immunol.* **106**, 267–71 (2000).

8. P. Kimmage, *Rough Ride.* (Yellow Jersey Press, London, 2001).

9. R. Bouchard, A. R. Weber, J. D. Geiger, 'Informed decision-making on sympathomimetic use in sport and health'. *Clin. J .Sport Med.* **12**, 209–24 (2002).

10. J. Swart et al., 'Exercising with reserve: evidence that the central nervous system regulates prolonged exercise performance'. *Br. J. Sports Med.* **43**, 782–8 (2009).

11. P. B. Medawar, *The Art of the Soluble.* (Methuen, London, 1967).

12. C. E. Cooper, 'Shining Light on the Body' http://web.me.com/profchris-cooper/Welcome/Light_and_Health.html). *Accessed February* 28, 2011.

13. L. M. Burke, 'Caffeine and sports performance'. *Appl. Physiol. Nutr. Metab.* **33**, 1319–34 (2008).

14. D. Silkstone, *The Age*, 'Stop the caffeine, world drug chief tells AFL' (July 8 2010) http://www.theage.com.au/afl/afl-news/stop-the-caffeine-world-drug-chief-tells-afl-20100707-10olq.html *Accessed March* 4, 2011.

15. S. D. Mahajan et al., 'Therapeutic targeting of "DARPP-32": a key signaling molecule in the dopiminergic pathway for the treatment of opiate addiction'. *Int. Rev. Neurobiol.* **88**, 199–222 (2009).

16. M. H. Eskelinen, M. Kivipelto, 'Caffeine as a protective factor in dementia and Alzheimer's disease'. *J. Alzheimers Dis.* **20** Suppl 1, S167–174 (2010).

17. T. Tysome, 'Pills provide brain boost for academics,' *Times Higher Education*, 29 June 2007. http://www.timeshighereducation.co.uk/story.asp?storyCode = 209480§ioncode= 26.

18. United States District Court for the Eastern District of Pennsylvania:United States of America vs. Cephalon Inc. Guilty Plea Agreement (2008).

19. P. Gerrard, R. Malcolm, 'Mechanisms of modafinil: A review of current research'. *Neuropsychiatr. Dis. Treat.* **3**, 349–64 (2007).

20. M. Xia et al., 'Identification of compounds that potentiate CREB signaling as possible enhancers of long-term memory'. *Proc. Natl. Acad. Sci. U S A* **106**, 2412–17 (2009).

21. D. Repantis, P. Schlattmann, O. Laisney, I. Heuser, 'Modafinil and methylphenidate for neuroenhancement in healthy individuals: A systematic review'. *Pharmacol. Res.* **62**, 187–206 (2010).

22. P. H. Canter, E. Ernst, 'Ginkgo biloba is not a smart drug: an updated systematic review of randomised clinical trials testing the nootropic effects of G. biloba extracts in healthy people'. *Hum. Psychopharmacol.* **22**, 265–78 (2007).

23. B. Roelands et al., 'The effects of acute dopamine reuptake inhibition on performance'. *Med. Sci. Sports Exerc.* **40**, 879–85 (2008).

CHAPTER 8

1. 'Laker, England, gets 8 wickets for 2 runs', *The New York Times*, June 1 1950.

2. G. P. Beunen, M. A. Thomis, M. W. Peeters, 'Genetic Variation in Physical Performance'. *The Open Sports Sciences Journal* **3**, 77–80 (2010).

3. C. R. Darwin, *The origin of species by means of natural selection, or the preservation of favoured races in the struggle for life, 6th edition; with additions and corrections.*, (John Murray, London, 1872).

4. D. S. Gardner, 'Historical progression of racing performance in the Thoroughbred horse and man'. *Equine Vet. J.* **38**, 581–3 (2006).

5. M. W. Denny, 'Limits to running speed in dogs, horses and humans'. *J. Exp. Biol.* **211**, 3836–49 (2008).

6. R. J. Herrnstein, C. A. Murray, *The Bell Curve: intelligence and class structure in American life.* (Free Press., New York, 1994).

7. J. Entine, *Taboo: Why Black Athletes Dominate Sports and Why We're Afraid to Talk About It.* (PublicAffairs, New York, 1999).

8. J. Stone, C. I. Lynch, M. Sjomeling, J. M. Darley, 'Stereotype Threat Effects on Black and White Athletic Performance'. *Journal of Personality and Social Psychology* **77**, 1213–27 (1999).

9. R. Bannister, paper presented at the The British Association for the Advancement of Science, Newcastle, September 13 1995.

10. R. A. Scott, Y. P. Pitsiladis, 'Genotypes and distance running: clues from Africa'. *Sports Med*, **37**, 424–7 (2007).

11. C. Bouchard et al., 'The HERITAGE family study. Aims, design, and measurement protocol'. *Med. Sci. Sports Exerc.* **27**, 721–9 (1995).

12. www.nitrxgen.net/factorialcale.php *Accessed March 4, 2011.*

13. E. A. Ostrander, H. J. Huson, G. K. Ostrander, 'Genetics of athletic performance'. *Annu. Rev.Genomics Hum. Genet.* **10**, 407–29 (2009).

14. M. S. Bray et al., 'The human gene map for performance and health-related fitness phenotypes: the 2006–2007 update'. *Med. Sci. Sports Exerc.* **41**, 35–73 (2009).

15. J. A. Timmons et al., 'Using molecular classification to predict gains in maximal aerobic capacity following endurance exercise training in humans'. *J. Appl. Physiol.* **108**, 1487–96 (2010).

16. H. E. Montgomery et al., Human gene for physical performance. *Nature* **393**, 221–2 (1998).

17. A. Jones, H. E. Montgomery, D. R. Woods, 'Human performance: a role for the ACE genotype?' *Exerc. Sport Sci. Rev.* **30**, 184–90 (2002).

18. L. Bahi et al., 'Does ACE inhibition enhance endurance performance and muscle energy metabolism in rats?' *J. Appl. Physiol.* **96**, 59–64 (2004).

19. N. Yang et al., 'ACTN3 genotype is associated with human elite athletic performance'. *Am. J.Hum. Genet.* **73**, 627–31 (2003).

20. A. K. Niemi, K. Majamaa, 'Mitochondrial DNA and ACTN3 genotypes in Finnish elite endurance and sprint athletes'. *Eur. J. Hum.Genet.* **13**, 965–9 (2005).

21. N. Yang, F. Garton, K. North, 'alpha-actinin-3 and performance'. *Med. Sport Sci.* 54, 88–101 (2009).

22. A. G. Williams, J. P. Folland, 'Similarity of polygenic profiles limits the potential for elite human physical performance'. *J. Physiol.* **586**, 113–21 (2008).

23. D. Fell, *Understanding the Control of Metabolism* (Portland Press, London, 1996).

24. R. Goldschmidt, *The Material Basis of Evolution (reissue).* (Yale University Press, New Haven, 1982).

25. L. Cox, 'Super Strong Kids May Hold Genetic Secrets' (April 2, 2009) http://abcnews.go.com/Health/MedicineCuttingEdge/story?id = 7231487 *Accessed April 5, 2011.*

26. H. Amthor et al., 'Lack of myostatin results in excessive muscle growth but impaired force generation'. *Proc. Natl. Acad. Sc. U S A* **104**, 1835–40 (2007).

27. P. Hakimi et al., 'Overexpression of the cytosolic form of phosphoenolpyruvate carboxykinase (GTP) in skeletal muscle repatterns energy metabolism in the mouse'. *J. Biol. Chem.* **282**, 32844–55 (2007).

28. E. R. Barton-Davis, D. I. Shoturma, A. Musaro, N. Rosenthal, H. L. Sweeney, 'Viral mediated expression of insulin-like growth factor I blocks the aging-related loss of skeletal muscle function'. *Proc. Natl. Acad. Sci. U S A* **95**, 15603–7 (1998).

29. Y. X. Wang et al., 'Regulation of muscle fiber type and running endurance by PPARdelta'. *PLoS Biol.* **2**, e294 (2004).

30. G. Gao et al., 'Erythropoietin gene therapy leads to autoimmune anemia in macaques'. *Blood* **103**, 3300–2 (2004).

31. S. D. Harridge, C. P. Velloso, 'Gene doping'. *Essays Biochem* 44, 125–38 (2008).

32. P. Schjerling, 'Gene doping'. *Scand. J.Med. Sci. Sports* **18**, 121–2 (2008).

33. V. M. Rivera et al., 'Long-term pharmacologically regulated expression of erythropoietin in primates following AAV-mediated gene transfer'. *Blood* **105**, 1424–30 (2005).

CHAPTER 9

1. P. Dimeo, *A history of drug use in sport 1876–1976: beyond good and evil.* (Routledge, Abingdon, 2007).

2. J. S. Mill, *On Liberty*. (1859).

3. O. W. H. Holmes Jr, Charles T. Schenck v. United States, Supreme Court of the United States 249 US 247. (1919)

4. 'Drugs and the Law: Report of the independent inquiry into The Misuse of Drugs Act 1971'. *The Police Foundation* (1999).

5. *The History of Herodotus Book IV*, translated by G. Rawlinson http://classics.mit.edu/Herodotus/history.html. (1859).

6. A. Huxley, *The Doors of Perception: And Heaven and Hell.* (Vintage Classics, London, 2008).

7. United States District Court U.S. v. BOYLL 774 F.Supp. 133 D.N.M. (1991).

8. J. Stevens, *Storming Heaven: LSD and the American Dream.* (Grove Press, New York, 1998).

9. J. A. Miron, J. Zwiebel, 'Alcohol Consumption During Prohibition'. *The American Economic Review* **81**, 242–7 (1991).

10. 'Drugs—facing facts. The report of the RSA Commission on Illegal Drugs, Communities and Public Policy'. *The Royal Society for the encouragement of Arts, Manufactures & Commerce* (2007).

11. D. J. Nutt, 'Equasy—an overlooked addiction with implications for the current debate on drug harms'. *J. Psychopharmacol.* **23**, 3–5 (2009).

12. Letter from Professor Nutt to House of Common Science and Technology Committee. London (2009).

13. B. Houlihan, *Dying to Win*, 2nd edition (Chapter 5). (Council of Europe Publishing, 2002).

14. I. Waddington, A. Smith, *An Introduction to Drugs in Sport: Addicted to winning.* (Routledge, Abingdon, 2009).

15. 'National Center for Catastrophic Sport Injury Research Data Tables, Annual Survey of Football Injury Research 1931–2007' http://www.unc.edu/depts/nccsi/FootballInjuryData.htm.

16. R. D. Hawkins, C. W. Fuller, 'A prospective epidemiological study of injuries in four English professional football clubs'. *Br. J. Sports Med.* **33**, 196–203 (1999).

17. 'Human Enhancement Technologies in Sport'. *House of Commons Science and Technology Committee*, London (2007).

18. G. Lineker, *Daily Mirror*, 21 November, 1994.

19. I. Botham, *The Botham Report* p. 236–7. (Collins Willow, London, 1997).

20. D. Powell, 'Blessed be the pacemakers, for they got away with it,' *The Times*, May 5, 2004. www.timesonline.co.uk/article/0,,13849-1098019,00.html.

21. J. Savulescu, N. Bostrom, Eds., *Human Enhancement*, (Oxford University Press, Oxford, 2009).

22. M. J. Sandel, 'The case against perfection: what's wrong with designer children, bionic athletes and genetic enhancement?' in *Human Enhancement*, J. Savulescu, N. Bostrom, Eds. (Oxford University Press, Oxford, 2009).

23. R. Kurzweil, *The Singularity is Near* p. 374. (Gerald Duckworth & Co. Ltd, London, 2005).

24. A. J. Schneider, R. B. Butcher, 'The mesalliance of the olympic ideal and doping: why they married, and why they should divorce' in *Sport…The third millennium: Proceedings of the International Symposium, Quebec City, Canada*, F. Landry, Ed. (Presses de l'Universite Laval, 1991), pp. 494–501.

25. J. Savulescu, B. Foddy, M. Clayton, 'Why we should allow performance enhancing drugs in sport'. *Br. J. Sports Med.* **38**, 666–70 (2004).

26. M. Burke, T. Roberts, 'Drugs in Sport: An issue of morality or sentimentality?' *J. Philos. Sport* **24**, 99–113 (1994).

27. I. Waddington, *Sport, Health and Drugs.* (Spon Press, London, 2000).

28. N. Bostrom, A. Sandberg, 'The wisdom of nature: an evolutionary heuristic for human enhancement', in *Human Enhancement*, J. Savulescu, N. Bostrom, eds. (Oxford University Press, Oxford, 2009).

29. R. M. Winslow et al., 'Different hematologic responses to hypoxia in Sherpas and Quechua Indians'. *J. Appl. Physiol.* **66**, 1561–9 (1989).

CHAPTER 10

1. R. K. Leute, E. F. Ullman, A. Goldstein, L. A. Herzenberg, 'Spin immunoassay technique for determination of morphine'. *Nat. New Biol.* **236**, 93–4 (1972).

2. F. Lasne, J. de Ceaurriz, 'Recombinant erythropoietin in urine'. *Nature* **405**, 635 (2000).

3. Z. Wu, M. Bidlingmaier, R. Dall, C. J. Strasburger, 'Detection of doping with human growth hormone'. *Lancet* **353**, 895 (1999).

4. P. A. Arndt, B. M. Kumpel, 'Blood doping in athletes—detection of allogeneic blood transfusions by flow cytofluorometry'. *Am. J. Hematol.* **83**, 657–67 (2008).

5. T. Beiter et al., 'Direct and long-term detection of gene doping in conventional blood samples'. *Gene Ther.* **18**, 225–31 (2011).

6. *Golf Today*, Nick Faldo speaks about drugs in golf (December 2004) http://www.golftoday.co.uk/news/yeartodate/news04/faldo6.html *Accessed March 10, 2011.*

7. 'Player: drugs in golf is a fact,' *The Guardian*, July 18 2007. http://www.guardian.co.uk/sport/2007/jul/18/golf.theopen20074.

8. M. S. Schmidt, 'Baseball Using Minor Leagues for a Drug Test,' *The New York Times*, July 22 2010. http://www.nytimes.com/2010/07/23/sports/baseball/23doping.html.

9. 'Der uneffektive Kampf gegen das Doping,' *Badische Zeitung*, Novembr 11 2010. http://www.badische-zeitung.de/sportpolitik/der-uneffektive-kampf-gegen-das-doping--37605657.html.

10. H. Striegel, R. Ulrich, P. Simon, 'Randomized response estimates for doping and illicit drug use in elite athletes'. *Drug Alcohol Depend.* 106, 230–2 (2010).

11. United States District Court for the Northern District of California: United States of America vs. Trevor Graham. Day 4 transcript (2008).

12. C. Lundby, N. J. Achman-Andersen, J. J. Thomsen, A. M. Norgaard, P. Robach, 'Testing for recombinant human erythropoietin in urine: problems associated with current anti-doping testing'. *J .Appl. Physiol.* **105**, 417–19 (2008).

13. M. Beullens, J. R. Delanghe, M. Bollen, 'False-positive detection of recombinant human erythropoietin in urine following strenuous physical exercise'. *Blood* **107**, 4711–13 (2006).

14. F. Lasne, 'No doubt about the validity of the urine test for detection of recombinant human erythropoietin'. *Blood* **108**, 1778–9; author reply 1779–1780 (2006).

15. D. Catlin, G. Green, M. Sekera, P. Scott, B. Starcevic, 'False-positive Epo test concerns unfounded'. *Blood* **108**, 1778; author reply 1779–1780 (2006).

16. A. T. Kicman et al., 'Candida albicans in urine can produce testosterone: impact on the testosterone/epitestosterone sports drug test'. *Clin. Chem.* **48**, 1799–17801 (2002).

17. P. Watson, C. Judkins, E. Houghton, C. Russell, R. J. Maughan, 'Urinary nandrolone metabolite detection after ingestion of a nandrolone precursor'. *Med Sci Sports Exerc* **41**, 766–72 (2009).

18. K. Sharpe, M. J. Ashenden, Y. O. Schumacher, 'A third generation approach to detect erythropoietin abuse in athletes'. *Haematologica* **91**, 356–63 (2006).

19. Y. Bhambhani et al., 'Boosting in athletes with high-level spinal cord injury: knowledge, incidence and attitudes of athletes in paralympic sport'. *Disabil. Rehabil.* **32**, 2172–90 (2010).

20. C. J. Gore et al., 'Second-generation blood tests to detect erythropoietin abuse by athletes'. *Haematologica* **88**, 333–44 (2003).

21. I. Erotokritou-Mulligan et al., 'The use of growth hormone (GH)-dependent markers in the detection of GH abuse in sport: Physiological intra-individual variation of IGF-I, type 3 pro-collagen (P-III-P) and the GH-2000 detection score'. *Clin. Endocrinol. (Oxf)* **72**, 520–6 (2010).

CONCLUDING REMARKS

1. A. J. Schneider, R. B. Butcher, 'An ethical analysis of drug testing', in *Doping in Elite Sport: the Politics of Drugs in the Olympic Movement*, W. Wilson, E. Derse, eds. (Human Kinetics Publishers, Inc., Champaign, 2001), pp. 129–52.

Further reading

The following books were particularly helpful during the writing process.

C. E. Cooper, R. Beneke, eds., *Drugs and Ergogenic Aids to Improve Sport Performance, Essays in Biochemistry* vol. 44 (Portland Press, London, 2008).

Covers much of the same field as this book but in more biochemical depth.

B. Houlihan, *Dying to Win*, 2nd edn. (Council of Europe Publishing, Strasbourg, 2002).

Very good on the ethical issues discussed in Chapter 9.

A. Jeukendrup, M. Gleeson, *Sport Nutrition: An Introduction to Energy Production and Performance.* (Human Kinetics, Champaign, Il., 2004).

Written by two of the top exercise biochemists in the UK. More comprehensive and up-to-date than Newsholme if less quirky in style.

W. D. McArdle, F. I. Katch, V. I. Katch, *Exercise Physiology: Energy nutrition and human performance.* 4th edn. (Williams and Wilkins, Philadelphia, 1996).

Good background reading—more recent editions are available.

E. Newsholme, A. Leech, G. Duester, *Keep on Running: Science of Training and Performance.* (Wiley-Blackwell, Oxford, 1994).

A great crossover book, aimed at runners, but with enough biochemistry to interest the afficianado. One of the books that got me interested in sports science. There is even space for some recipes!

J. Savulescu, J., N. Bostrom, N. (eds.). *Human Enhancement,* (Oxford University Press, Oxford, 2009).

A selection of essays about the ethics of all forms of human enhancement putting drugs in sport in the context of wider societal questions.

I. Waddington, A. Smith, *An Introduction to Drugs in Sport: Addicted to winning.* (Routledge, Abingdon, 2009).

One of the more accessible sociology books.

J. M. Wrigglesworth, *Energy and Life.* (Tayor and Francis, Basingstoke, 1997).

The antidote to loose and wooly thinking about energy in biology, written by one of my scientific mentors.

INDEX